Dog's Best Friend

ALSO BY SIMON GARFIELD

Expensive Habits

The End of Innocence

The Wrestling

The Nation's Favourite

Mauve

The Last Journey of William Huskisson

Our Hidden Lives (ed.)

We Are at War (ed.)

Private Battles (ed.)

The Error World

Mini

Exposure

Just My Type

On the Map

To the Letter

My Dear Bessie (ed.)

A Notable Woman (ed.)

Timekeepers

In Miniature

Dog's Best Friend

THE STORY OF AN UNBREAKABLE BOND

Simon Garfield

wm
WILLIAM MORROW
An Imprint of HarperCollins*Publishers*

Pages 293–294 constitute a continuation of this copyright page.

"Man and Dog" by Siegfried Sassoon is copyright Siegfried Sassoon, and is reprinted by kind permission of the Estate of George Sassoon.

HarperCollins books may be purchased for educational, business, or sales promotional use. For information, please email the Special Markets Department at SPsales@harpercollins.com.

Originally published in the United Kingdom in 2020 by Weidenfeld & Nicolson.

A hardcover edition of this book was published in 2020 by William Morrow, an imprint of HarperCollins Publishers.

FIRST WILLIAM MORROW PAPERBACK EDITION PUBLISHED 2021.

Designed by Elina Cohen
Image on pages iv, ix, and xiv from Shutterstock by Barpe

Library of Congress Cataloging-in-Publication Data has been applied for.

ISBN 978-0-06-305225-3

21 22 23 24 25 LSC 10 9 8 7 6 5 4 3 2 1

FOR ALL THE DOGS WE LOVE.

If you don't own a dog, at least one, there is not necessarily anything wrong with you, but there may be something wrong with your life.

—ROGER CARAS

The dog has seldom pulled man up to his level of sagacity, but man has frequently dragged the dog down to his.

—JAMES THURBER

The fact that a dog can smell things a person can't doesn't make him a genius; it just makes him a dog.

—TEMPLE GRANDIN

CONTENTS

INTRODUCTION: THE DOGNESS OF DOGS

Why is he here?

Why is my dog lying at my feet in the shape of a croissant as I write this? How have I come to cherish his warm but lightly offensive pungency? How has his fish breath become a topic of humor when friends call round for dinner? Why do I shell out more than a thousand dollars each year to pay for his insurance? And why do I love him so much?

Ludo is not a special dog. He's just another Labrador retriever, one of approximately 500,000 in the U.K. (he'd be one in a million in the United States, the most popular breed in both countries). Ludo has a lot in common with all these dogs. He loves to play ball; obviously he's an expert retriever. He could eat all the food in the universe and leave nothing for the other dogs. He is prone to hip dysplasia. He looks particularly attractive on a plush bed in a centrally heated house very far from the Newfoundland home of his ancestors.

But of course Ludo is a unique animal to me and the rest

of his human family. He is now an elderly gentleman aged twelve and a half, and we would do almost anything to ensure his continued happiness. We willingly get drenched as he tries to detect every smell in the park. We schedule our days around his needs—his mealtimes, his walks, the delivery of his lifesaving medication (he has epilepsy, poor love). We spend a bizarrely large amount of our disposable income on him, and he never sends a card of thanks.* When he's not with us for a few days (when our children take him for a weekend, say) then the house feels extraordinarily empty. I feel so fortunate to know him. Goodness knows how we'll cope when he dies.

This weekend I will visit Discover Dogs at an exhibition center in east London to watch dogs perform agility and obedience tests in a ring, and I will have the chance to meet two hundred different breeds, some of whom would fit in my bag and some who would have trouble fitting in my car. I'll also have the opportunity to buy a vast amount of dog-related paraphernalia and crap, the majority of which is not of course for dogs but humans, including oil paintings, clothing and dog-related homeware (with slogans such as "If I can't bring my dog I'm not coming," "Dogs make me happy, you not so much" and "I'd rather be walking my Schnauzer"). To compensate for the fact that family pets are not allowed at this event, the following Friday Ludo will attend a screening of *Rocketman* at the Exhibit cinema

---

* If you're reading this at a point where you're thinking of getting your first dog and consider a purchase price of $1,000 a little dear, then all I can say about the costs to come—vets, food, dog-sitting, accoutrements both essential and superfluous—is ha ha ha.

in Balham, south London. Although no particular fan of Elton John (he likes listening to anything, really, so long as it doesn't sound like a vacuum cleaner), Ludo will enjoy his own seat next to mine, with a blanket and "pupcorn" treats. All the dogs at this screening will gain free admission "in exchange for cuddles with the team," and the lights stay a little higher during the film so as not to distress them.

How did we get here, to the point where the dog is top dog? How did we arrive at the moment when a dog goes to the movies? How and when did we realize that dogs would assist humans not only in hunting, but also in bomb disposal and cancer detection? With what degree of quiet acquiescence did humans roll over and accept that our domestic lives—our work hours, the cleanliness of our rugs, our holiday choices—were henceforth to be determined by the demands of an animal that used to live outside and fend for itself? When and why did the sofa replace the scavenging?

This book examines how this strongest of interdependent bonds has manifested itself over the centuries, and how it has transformed so many millions of lives, human and canine. If it is at least partially true, as Nietzsche claims, that "the world exists through the understanding of dogs," then perhaps it is also partially true that a study of dogs may provide a valuable insight into ourselves.

WHY IS HE HERE?

Why is this man doing something that involves a repeated tapping noise and the occasional loving sigh? How many hot drinks can he make to interrupt this tapping? Why is his time-keeping so bad when it comes to my luncheon? Why can't this so-called memory foam bed he bought me remember how I curled up so snugly last night? Why do I feel so fortunate to know him?

The anthropomorphism of dogs is not a new phenomenon. I have a photo on my desk of a black Labrador from the nineteenth century dressed as a lord in a suit and top hat (and smoking a pipe). Talking dogs have been a mainstay of film almost from the birth of talking movies. But the collusion of dog and human has never been so abundant, imaginative and unnerving as it is today. The nature of our bond—our commitment to each other—appears to have deepened markedly in the past fifty years, not least because our scientific understanding of the dog has been enabled by advances in genetics, and our sociological interpretation of a dog's behavior has led to more avenues for joint engagement. Like dancers emboldened by drink and tenacity, we are entwined with our best friends in an ecstatic embrace.

Such passion does not always end well, alas. Alongside my Victorian lord I have a photo of a dog in a flat Kangol cap and glasses who looks like Samuel L. Jackson. On my computer I have pictures of dogs reading, sailing and riding bicycles. I know there is something morally wrong with these images, but I find it hard to resist adding more to the folder, given their wholly irresistible paws-to-the-floor adorableness.

Every week I get an email from the magazine *Bark* with the subject line "Smiling Dogs." Each message contains at least two pictures of beautiful grinning hounds, most recently Baxter ("Baxter has a bubbly personality, loves food, lounging in the sun, hiking outdoors, and cuddling") and Chad ("This handsome boy might come across as a little aloof at first but that's what makes him mysterious and charming!"). Appealing as these dogs are, they are not, of course, actually grinning. But the people at *Bark* know well that the photogenic often get a head start: most of the dogs in the emails are looking for new homes after a harsh beginning.

The names we give our dogs are increasingly names we would give to our children. For every old Fido we have a new Florence; for every old Major we have a new Max. This was not the case thirty years ago. Today the new names are the names of human heroes. Nelson is still popular; soon we will see a lot of Gretas. You have a female dog called Taylor, you will have a male one called Swift. Lawyers like to call their dogs Shyster, and architects favor Zaha, and there are an awful lot of young Fleabags in the parks these days. Only in rap music does it work the other way: Snoop Dogg, Phife Dawg, Nate Dogg, Bow Wow.

We increasingly use dogs to describe ourselves. A tough radio interviewer is a Rottweiler, a soft one a poodle (or a puppy). Friendly, faithful characters in novels are cuddly Labradors. Venal men in the city are pit bulls. A person who won't let go fights like a terrier, while a detective pursues her prey like a bloodhound. You get the idea. You get the idea because you are as fleet as a whippet and as smart as a sheepdog.

We have long used our canine friends to describe our actions and emotions. After working like a dog we are dog-tired. We get drunk as a skunk but we drink the hair of the dog. Books containing doggerel get dog-eared. We root for the underdog, we bark up the wrong tree and then we're in the doghouse. A depression is a black dog, and we'll sport a hangdog countenance. A dog's breakfast is followed by a dog's dinner, but the dog ate my homework so I've gone to the dogs. And we have sex in a position so popular among dogs that they have officially trademarked the style.

I AM FINISHING THIS book during the virus-haunted days of April 2020, and Ludo is the only presence in our house not looking anxious. Instead he is exhausted. It has already become a cliché to observe that the pandemic has been perversely kind to the domestic dog: they are seldom home alone now, and they are walked almost more than they can bear. Friends and neighbors want to borrow him: if you have a dog, you have a reason to be out. Rescue shelters report a surge in inquiries. The venue which only a few months earlier hosted Discover Dogs is now a four-thousand-bed hospital. Social media is awash with COVID-19 dog videos and cartoons. Dog owners who live alone are more grateful than ever for the company and comfort. But there are additional worries too: today more than ever I do not want Ludo to be unwell; his regular dry food from Germany is in short supply; those scented poo bags are very hard to open without first licking your fingers to separate the opening.

Even if you have never owned a dog, and even if you

have only watched the Westminster Dog Show on television, you will know that our relationship with dogs is a rich, diverse, perplexing and complicated one—as rich, diverse, perplexing and complicated indeed as the relationship we have with other humans. Dogs are increasingly not just part of the home but part of the family, the closest connection we dare have with a species not our own.

In many ways dogs have become an extension of ourselves. Albert Einstein once observed that Chico, his wire-haired fox terrier, was possessed of both great intelligence and an ability to hold a grudge. "He feels sorry for me because I receive so much mail; that's why he tries to bite the mailman." This approach—only the social scientists persistently call it anthropomorphism; dog lovers tend to regard it as entirely acceptable behavior—is widely frowned upon by most animal behaviorists as *in*humane. But still we do it. In fact, we now do it with such conviction and sense of normality that not to treat our dogs to a diet involving turmeric may come to seem like neglect.

When the Exhibit cinema began its screenings for dogs and their owners in 2017, the films had a dog-related theme—*Lady and the Tramp, Isle of Dogs*—but more recently the dogs have watched regular movies: *If Beale Street Could Talk, The Big Sick, Can You Ever Forgive Me?* When the movie is over, the modern dog retains its Hollywood glamour. We wrap them in furs and bejeweled collars, make stars of them on Instagram.*

---

* Maybe the experience of going to the cinema with your dog isn't so new after all. Here's a joke from the 1990s: "The last time I went to the cinema there was a woman there with her dog. The dog laughed all through the

This book is primarily a celebration of dogs in all their intelligence, curiosity, beauty and loyalty. Would dogs write a similar appreciation of humans? I wonder. We shall indulge in dog stories reassuring and absurd, warming and alarming, funny and severe. But the book will also pose a few difficult questions about the way humans now treat our great canid friends, and where this may be leading us. It questions, for example, whether the natural love we have for our beloved pets isn't spilling over into disrespect, and our love of variety and novelty into exploitation. All the breeders I talked to are worried about the future. Have we forgotten where dogs came from and how they used to live? Do we always provide the best life for them, as opposed to the best life for us? And are we in danger of losing what the canine psychologist Alexandra Horowitz has called "the dogness of dogs"?

At the core of this book is one leading question: How did we get from hunting with the Eurasian wolf (among other species of *Canis lupus*), to buying an electrically heated day-bed for the Cavalier King Charles Spaniel (among other species of *Canis lupus familiaris*)? The journey is cultural and scientific, and takes us across the globe.

En route I will attempt to explain the origin of the cheagle (the Chihuahua and beagle cross) and the chi-weenie (Chihuahua and dachshund), and the very notion of the designer dog. I will recount the sequencing of the

---

film, but when the hero died at the end he was in tears. When it was over I went up to the woman and said I thought her dog was extraordinary, and how I'd never before seen a dog enjoy a film so much. 'I'm as amazed as you are,' she said, 'because he really hated the book.'"

first complete dog genome, and will consider the most significant recent experiments and theories from the science journals. I'll ask whether Charles Darwin shouldn't be as well known for his work on dogs as he is for his work on evolution, and examine why Charles Dickens wanted to buy a gun to shoot dogs at random. I will explore a secluded dog cemetery and other ways we choose to memorialize our beloved pets. I will also try to understand why prints of dogs playing poker were once bestsellers, and why, if you haven't yet seen it, you need to search YouTube for "Ultimate Dog Tease," a video in which a dog called Clark is repeatedly disappointed when his owner refuses to give him the bacon treats he so obviously deserves, and which has been watched more than 200 million times.

But I am not a psychologist or ethologist, much less a geneticist, so I've sought wise guidance from specialized minds in these fields. My own explorations are journalistic and evidential, and for the best evidence I have relied on a sequence of dogs who have sat somewhere near my desk for thirty years: a basset hound called Gus, a yellow Labrador retriever named Chewy and my black Labrador, Ludo. You cannot know a well-mannered dog for any length of time—more than, say, about an hour—and not wonder a little about what he or she is thinking, what makes him or her fearful or happy, and how the two of you may have fun together. (The book is skewed toward the positive. There are many vicious dogs in the world—I was once bitten by one as I rode my bike back from school, a German shepherd: tetanus jab for me, angry letter from my solicitor father for its owner—but I have decided to focus on the harmonious side of our relationship, which is happily the dominant one.)

A dog resides superbly within what the German biologist Jakob Johann von Uexküll called its own self-centered world, or *Umwelt.* Or, as the primatologist Frans de Waal put it in the title of his book *Are We Smart Enough to Know How Smart Animals Are?* if a dog cannot fully comprehend systems of time and money, it is not because he is unintelligent; it is because these things are not significant components of his world.

The average dog brain is about one-third the size of an average human brain. But the dog nose has more than 200 million smell-sensitive receptors, compared to 5 million in a human nose, and these suggest a quite different set of priorities. About a third of the dog's brain mass is devoted to olfactory duties, compared to 5 percent in humans. I can't help but notice how my own dog, with his proud leathery snout, views the world around him. His exacting sense of smell makes him a very good judge not only of his environment and other dogs, but also of people: he can judge who may be frightened of dogs and keep away; he remembers who has paid particular attention to him in the past and will make sure to greet them with gladness in his heart and a special toy in his mouth; and he knows when his human companions are low and need comforting. I sometimes wonder whether we are treating him and his many friends with a similar level of sagacity and respect.

One of the many things that attracts us to a puppy—beyond their all-around damn helpless cuteness—is their inquisitiveness. Puppies like poking around in things, any things. This inquisitiveness matures, but it doesn't depart: older dogs hear an irregular noise, they still want to investigate. Perhaps one could regard this book as a dog discovering the world around it: irregular noises, a rapidly changing environment and an increasingly large amount of attention from complete strangers. These strangers are us, also acting at our most puppyish, discovering with increasingly forensic precision just what it is that makes a dog a dog, and makes them such mutually enriching companions. And we are strangers only to ourselves: as dog owners and dog lovers, we are part of a huge community, and the bond we have with our dog is something that binds us equally to millions of other humans, a shared humanity.

Where best to begin our historical survey of this relationship? Perhaps with the visual evidence, with dogs at their most irrepressible, and humans at their most indebted.

1.

The Indelible Image

For a few days in February 2019, Protein Studios in Shoreditch, London, decided to hang some of its pictures at dog-eye level. The show featured "famous dogs of yesteryear" such as the Queen's corgis, Laika the Soviet space dog and Petra the *Blue Peter* dog sitting at a typewriter answering her fan mail. There were also photographs of "hero dogs" who had wheels where their hind legs had been, and pictures of the most photogenic canine influencers on Instagram. The exhibition, which was called *The National Paw-Trait Gallery* (the hyphen was probably unnecessary), was a launch promotion for the World's Most Amazing Dog competition run by Facebook, although the definition of "amazing" wasn't entirely clear, and when a nine-year-old Chihuahua from Mexico named Toshiro Flores was declared the winner, he seemed as surprised as anyone.

Dogs have been portrayed in public galleries since humans painted on the walls of caves, and every private gallery owner since the Renaissance has acknowledged one universal truth: you put a dog on a wall and people will come panting. It may also be true that if you put enough dogs on enough walls you can construct a compelling survey of the human–dog relationship over thousands of years.

What can we glean from other recent shows? In 2013, an exhibition at The Gallery on the Corner in Battersea, London, was selling art in aid of Battersea Dogs & Cats Home that was actually created by dogs: leathery snouts pushed a bowl of food around the floor, the bowl had a brush attached to it, and the floor was lined with paper. Dogs were welcome to view as well as participate in this show, and they were especially welcome if they were in the mood to buy.*

In July 2019 Southwark Park Galleries in southeast London held a similar show. Billed as "Contemporary Art. Chosen by Dogs. For Dogs and Humans," it was a case of canine-owning gallerists and critics choosing their favorite dog-related creations. There was work from the artists Martin Creed, Joan Jonas, David Shrigley and Lucian Freud. The etchings, oils, films and film stills showing many dogs in many different situations didn't appear to have anything in common beyond the gimmickry of their

* A couple of the dogs displayed natural talent, but in my opinion most of them still had a long way to go. Some of their work lacked even the most basic grounding in art theory, and, frankly, much of it just looked unfinished. Human painters often find it very hard to walk away from their canvasses and declare them done, but not these dog painters—these painters couldn't get to their food rewards fast enough.

curation, but on second glance a commonality emerged: they were all adorable. The love we have for dogs is nowhere better displayed than on canvas or photograph, nor our dependence, nor their purpose.

The works at Southwark Park Galleries are part of a noble pantheon. A stroll around any major gallery provides a canine period and scenario for every mood, and one may track how the human–dog relationship changes over the centuries. One begins in the fifteenth century with the dog as a partner in the hunt and a symbol of solidity and—the animal as prized aristocratic possession—one ends with dogs in fancy hats garnering millions of online likes. Dogs on Instagram are no less significant than the hunting scene, for both are images of delight. The dogs have changed slightly—in build, in prominence, certainly in variety—but their importance to the image and image-maker is consistent.

A walk through the public rooms and storage areas of the National Gallery in London will yield some two hundred paintings of dogs, most of them seemingly incidental. But look again: so many of the dogs are in charge of the canvas, subtly dominating the image just as they have subtly charmed their creator. Their prominence confirms their importance; even their seemingly incidental presence—say the tiny bourgeois Brussels griffon at the foot of Jan van Eyck's *Arnolfini Portrait* of 1434—often carries a weighty message, in this case fidelity and pride. Here is *Christ Nailed to the Cross* by Gerard David from about 1481, the almost naked central figure stretched prostrate diagonally across the canvas, and in the foreground a small, almost hairless dog sniffing a skull; here was destiny, and a curiosity about what becomes of us. Here is *Joseph Greenway* by Jens Juel (1778), a dandy in woodland with his hunting dog looking up at him with respect and a little fear, much as the crew on Greenway's trading ships must have done. Here is Canaletto's subtle remodeling of Piazza San Marco in Venice in 1756, a scruffy terrier-like mutt at the feet of two noblemen: the dog looks to be awaiting pastry scraps from Caffè Florian to its right; any dog owner will recognize the hope.

These are incidental dogs, dogs who put their noses in. Elsewhere, in galleries worldwide, the dogs move to the foreground and there is a hound for every emotion and mood. You want aristocratic lording? Look at Gustave Courbet's *Greyhounds of the Comte de Choiseul* (1866). Dog as protector? Try Jeanne-Elisabeth Chaudet's *An Infant Sleeping in a Crib Under the Watch of a Courageous Dog Which Has Just Killed an Enormous Viper* (1801). Stone-cold cuteness? Philip Reinagle's *Portrait of an Extraordinary Musical Dog* (1805) with paws on a keyboard and an expression of "just practicing!" on its little spaniel face. You'll find utter contempt in the dachshund urinating onto the central figure in *Match Seller* by Otto Dix (1920), and utter radiance in the many thick-lined, straight-backed hip-hop mutts created by Keith Haring. Again, one looks for commonality and comes up wanting. Why should they have anything but fur to share between them? But soon enough a thread emerges: the dogs draw us in with their warmth, their consoling presence, their very dogness. However central or slender their presence on the canvas, all the pictures would seem incomplete without them. And painful too, like a cut along a finger joint.

I once visited David Hockney at his studio in Los Angeles, and inevitably we talked about his beloved dachshunds, Stanley and Boodgie. They were tricky sitters, he said, their attention easily diverted by visitors and any activity in the kitchen. He concluded that they were not hugely interested in art.

Hockney has long outlived both of them, just as he has outlived most of his closest human companions. In their memory he created a dog wall at his house in Los Angeles, a gloriously pungent spread of warmth and affection: curled dogs, dogs on their backs, dogs nuzzling each other on their tender beds, dogs with their snouts dreaming off the edge of a cushion. There is a particularly moving photograph of Hockney slumped in a striped armchair in front of this huge wall, an arm curled proudly around each dog, with more than forty paintings of Stanley and Boodgie behind him. "I make no apologies for the apparent subject matter," he wrote in his introduction to a book of his dog portraits. "These two little dear creatures are my friends. They are intelligent, loving, comical and often bored. They watch me work. I notice the warm shapes they make together, their sadness and delights." He explained that in a world of sadness he wanted desperately to paint something loving. "The subject wasn't dogs but my *love* of the little creatures." This is, of course, what we perceive in almost all dog-related art.

Further evidence of the almost suffocating affections that humans lavish on their dogs will be found at the Kennel Club in Mayfair and the American Kennel Club's Museum of the Dog in Manhattan, repositories of the biggest and most ornate collection of canine art in the world. Here is the dog as hero, as superb specimen, as blood-sports warrior. The London collection has a great many championship trophies and certificates from more than a century of dog shows, and a vast array of photographs portraying proud, devoted humans alongside proud, exhausted dogs. There are royals, commoners and a large number of eccentric people in tweed and brogues whose elevator didn't ride all the way to the top. The collection also documents eloquently the various and changing roles of the dog in nineteenth-century England, from its many coursing and hunting scenes to an engraving of Billy, the celebrated ratcatcher of 1823 (possibly a terrier, certainly a champion: Billy is shown in a ring killing one hundred rats in just over five minutes). The most poignant, and the most narratively succinct portrayal of how the working dog was being overtaken in our affections by the domestic pet came in 1860 from Richard Ansdell's oil painting *Buy a Dog Ma'am?* Heavily influenced by Sir Edwin Landseer, and originally shown at the Royal Academy, it shows a hardened-looking but rather uncaring man by a large pillar in what is apparently a market area in a city. In one hand he holds up a white toy dog with a red bow around its neck (perhaps a poodle mixed with a pug), while under the other he guards what may be a spaniel. By his feet stand two distinctly dejected larger working dogs, and the message is clear: their time in the field has passed, and they are no longer fit for duty.

The Manhattan collection moved to its new home on Park Avenue at the beginning of 2019, after many quiet years in the suburbs in St. Louis. It contains the sort of amusing doggy knickknacks of which any great-aunt would be proud (china figurines, pewter hunting trophies), but also some proper "hey-wow" exhibits, including a photofit screen that determines, should you ever transform into a dog, what sort of dog that will be (the computer coding goes by your looks rather than your temperament).

Pictorially the Museum of the Dog has all the classics, including Maud Earl's *Silent Sorrow* (Edward VII's dog Caesar miserably mourning his loss on the side of an armchair in 1910) and John Sargent Noble's *Pug and Terrier* of 1875, the terrier tied up and forlorn with a begging bowl around its neck marked "charity," while the well-fed pug stands on a step above him looking sorrowful at the injustice of it all. But the most notorious picture, given a wall panel all to itself, is a painting by Christine Merrill of *Millie on the South Lawn* of the White House. Millie was an English springer spaniel once owned by George H. W. Bush and Barbara Bush, and there she sits next to a red ball, taking up almost the whole canvas, the White House and its fountain behind her, practically an afterthought. Millie looks as if she's in charge; she looks as if her owners love her. The portrait is accompanied by a letter from Barbara Bush marking the opening of the dog museum in St. Louis in 1990: "Dogs have enriched our civilization," she writes, "and woven themselves into our hearts and families through the ages . . ." (Donald Trump is the first American president in more than a century not to own a dog in the White House.)*

The museum's most arresting item is not a painting but a parachute, a canopy used by a canine hero of the Second World War. Many dogs flew crucial missions during the conflict, including Rob the parachuting English collie, who apparently jumped or was pushed from a plane more than twenty times while working behind enemy lines for the SAS during the North African campaign, and at war's end was awarded the Dickin Medal, the canine equivalent of the Victoria Cross, Britain's highest military honor.†

---

* At a rally in El Paso in February 2019, Trump praised the German shepherds sniffing out drugs on the Mexican border, but he said the prospect of owning a dog himself would be "phony." In her memoir, his first wife, Ivana, wrote of how her poodle, Chappy, would "bark at him territorially" whenever Trump drew near. His predecessors took a different approach. The Obamas' two adopted Portuguese water dogs, Sunny and Bo, were regular stress relievers, and in such high demand for photo opportunities that they needed their own official White House schedule. The memoir of Barbara Bush's Millie, written by Millie herself (obviously), outsold those of both her owner and her husband. Trump's use of the word "dog" is consistently derogatory. His former chief strategist Stephen K. Bannon was "dumped like a dog by almost everyone," while former presidential candidate Mitt Romney "choked like a dog" during his campaign. Trump regularly noted how his enemies had been "fired like a dog," whatever that meant. And when, in late October 2019, Trump triumphantly announced the death of the Islamic State leader Abu Bakr al-Baghdadi, he crowed that "he died like a dog." He was prouder of the role that a dog named Conan played in the death itself, reportedly pursuing al-Baghdadi down a tunnel as he detonated a suicide vest. The president called Conan, who was a Belgian Malinois, "a beautiful dog—a talented dog."

† What a star. Or at least he was until 2006, when a former member of the SAS suggested Rob's achievements were hugely exaggerated, and that he may never have jumped at all, the ruse simply dreamed up by Rob's adopted

But the hero commemorated in Park Avenue's Museum of the Dog is the famous Yorkshire terrier Smoky. The precise details of his exploits are, like those of Rob, hard to prove, but it's believed Smoky fought in the jungles of New Guinea and helped set up communication lines beneath an important airstrip. She was attached to the 5th Air Force, 26th Photo Reconnaissance Squadron, and although she didn't take photos, she was credited with twelve combat missions and won eight battle stars. According to the *Yorkshire Post*, which covered the story because Smoky's family was originally local, the dog's efforts saved the lives of more than 250 men and more than forty planes. But that wasn't enough for Smoky; Smoky wanted more.

When her owner, Bill Wynne, had a lay-up in hospital, Smoky came to sit on his bed. Soon other patients wanted her near for comfort too, and Smoky became an in-demand therapy dog. When Wynne landed in Australia, he and Smoky toured the hospitals and soon the patients appeared well again. But that still wasn't enough for Smoky (or for Wynne). Smoky jumped out of a plane wearing a parachute to beat four hundred other entrants for the title Best Mascot of the Southwest Pacific Area. She then became a local celebrity wherever she went, and was a minor hit in Hollywood, not quite at the Rin Tin Tin or Lassie level, but still something of a draw at supermarket openings and on cable television.

---

minder in the air force, who was keen to prevent his return to his original owner.

• • •

TO FULLY appreciate the longevity of our artistic relationship with dogs we need to go back to the living museum known as Pompeii. The semi-domesticated dog was once everywhere here, and a visitor to the ruins today may sense a tail disappearing around every corner. A few were buried beneath the hot ash, but hundreds more fled (with or without their owners) when the early warnings of Vesuvius rumbled down in A.D. 79. The most famous of the remainers guarded the entrance to the House of the Tragic Poet, in the northwest section of the city, an essential stop on every tour.

This dog is, alas, a mosaic. Perhaps you've seen him: snarling in the vestibule, visible from the street, black and white with a thin red collar, chained up for now but ready to pounce should you even think of entering without permission. The words beneath him may, given the ferocity of the dog's intentions, be two of the most superfluous in the Latin language: *Cave canem.*

But perhaps "Beware of the dog" is just a sign; perhaps the mosaic negates the need for a real snarling dog. I've heard it suggested that similar signs in Pompeii once warned people to beware not because the dog was protective and a biter, but because it might be one of those small whippet-style creatures that curl up so sweetly at your feet that they are easy to trip over and harm.

But the Tragic Poet did not own one of those. In fact, the Tragic Poet owned nothing but his tragic verse. The house only took on his name when it was excavated in 1824, the title inspired by a wall painting of what was once believed to be a recitation of woeful verse to a rapt audience (but was later reinterpreted as the dramatic delivery of an oracle). So we do not actually know who lived in a house like this, or who owned (or pretended to own) the threatening dog. But the house is framed by two shops, one of which may have sold rings and necklaces, suggesting the possibility of a jeweler protecting precious gems.

And then there is the lava dog, the dog that died as Vesuvius erupted. Tethered and unable to flee, convulsed on its back in pain, howling out for its owner and its former Pompeiian life of sunken baths and toga parties.

This famous image is not, as many believe at first sight, and I did too, a dog enrobed in hardened ash, like edible nonsense in a too-fancy restaurant. It is not even an actual relic from Pompeii. It is instead a cast of a dog formed in the 1870s by injecting plaster of Paris and a gluey solution into the space created when the original occupant of this space—that poor dog, a chain attached to its thick collar—decayed and rotted away in the many centuries following his hot death. The dog left behind a cavity around which pumice had solidified, and archaeologists and curators saw their opportunity. You inject into the cavity, you wait till your filler hardens, and you chip and scrape and smooth away at its cast, *et voilà*: an agonized dog speaking to us down the centuries.

Accounts differ as to who owned the dog and where he lived. Some have him as a guard dog, chained up in the forecourt of the house of the Roman general Marcus Vesonius Primus. But others, including Mary Beard, grant him more humble origins, as the dog of a "fuller," a laundryman and cloth worker. Whichever is true, you may now see the plaster hound forever gasping its last at the at the Antiquarium of Boscoreale on the outskirts of Pompeii. Obviously no dog would wish to die this way, suffocating and alone. What were the animal's final thoughts? Why did its owner not return to save it?

The good news is the dogs came back to Pompeii. In a dramatic reconstruction of early domestication, the wild dogs of Naples and surrounding areas discovered that tourists were where the food was. So at the end of the twentieth century they returned in large and scavenging numbers, to the point where tourists reported themselves overrun and a little intimidated, and the Italian culture ministry decided they should act. In 2009 they established a Herculean dog rescue mission, wherein many of the stray dogs of

Pompeii were photographed and put up for adoption on the (now defunct) website (C)Ave Canem. More than twenty dogs were welcomed into caring homes in the campaign's first six months, which was judged a modest start, given that the Italian anti-vivisection league estimated there were seventy thousand strays in the surrounding Campania region. The advances and advantages of neutering and microchipping have been slow to arrive in this beautiful and ruined part of the world, and dogs find that they once again have the run of the place.

2.

How Dogs Began

On July 11, 2017, the *Journal of Anthropological Archaeology* received an excited email from a woman named Dr. Maria Guagnin. The message summarized her recent research from two dusty digs in northwestern Saudi Arabia, and the accompanying photographs showed engravings on ancient rocks depicting 147 hunting scenes, including the pursuit and capture of lions, ibex, gazelles and horses. Dr. Guagnin dated the carvings somewhere between 8000 and 6000 B.C., and explained how much they revealed about human survival on this arid portion of the Arabian Peninsula. But they also showed something else: the earliest visual evidence of the human domestication of dogs.

Dr. Guagnin, who earned her Ph.D. from the University of Edinburgh and worked at the School of Archaeology at Oxford before moving to the Max Planck Institute for the Science of Human History in Jena, Germany, is a specialist in prehistoric human–animal relationships. Three months after her email, the journal decided to bypass the traditionally sedate schedule associated with academic publishing and release Dr. Guagnin's paper online. The reaction was initially skeptical, but then ecstatic: Melinda Zeder, an archaeozoologist at the Smithsonian Institute in Washington D.C., called the report "truly astounding," which is not the sort of phrase scientists usually employ. Some of the photographs were enhanced by computer imaging, making the drawings look as if they had been newly sketched in chalk just a day before, and showed dogs biting the stomach and neck of an ibex. Others showed clearly that many of the dogs were on a leash. Some were tied to a hunter's waist, which freed up the hunter to use a bow and arrow. In one engraving, an armed hunter is surrounded by thirteen dogs, all facing the same direction, presumably toward their prey.

The rock art came from the Shuwaymis and Jubbah regions, an area that has long been rich in archaeological finds. In Shuwaymis, 273 rock art panels showed 52 dogs, while at Jubbah there were 127 dogs on 1,131 panels. The dogs were all of one ancient breed, the Canaan, named after the region populated by the Phoenicians in 500 B.C. Dr. Guagnin deduced that this particular set were either brought to the region from the Levant or were a direct descendant from Arabian wolves, and she noted how they all had similar characteristics: pricked-up ears, short snouts, curled tails. Many of them also displayed large white patches on their chests, and smaller white markings on their shoulders, a classic sign of the breed, also plainly evident on the rock carvings. More romantically, the Canaan was known as the Bedouin sheepdog or the free-roaming Palestinian pariah dog.

The depictions of these animals rewrote the history books. Anthropologists have long agreed that the domestication of dogs had begun tens of thousands of years ago, but this Arabian rock art was now the earliest artistic proof.*

In Dr. Guagnin's report, submitted with two coworkers at the Max Planck Institute, she speculated on the important presence of the leash: she wondered whether some dogs had been singled out as particularly valuable scent dogs, or protected on account of their young or old age. Perhaps they were also tethered to protect their owners lest their prey turned against them, or perhaps to help haul meat back to camp. The unleashed dogs may have had particular attributes as attackers, and there are several scenes in which dogs appear to have surrounded their targets at the edge of cliff tops. In this way the engravings suggest a clear advancement of the use of dogs as trained individuals rather than wild packs. Humans have assigned dogs distinct tasks, and possibly names as well. From this point on, dogs and humans are going to get along.

---

* Previously, the first representation of the domesticated hunting dog appeared on pottery recovered in Iran almost two thousand years later, around the time humans had begun to switch from hand-to-mouth subsistence hunting to the herding of sheep, cattle and goats. The Chauvet cave in southern France contains a set of footprints imprinted over a 150-foot stretch of clay that has been widely interpreted as showing a child walking alongside a semi-domesticated wolf some 26,000 to 28,000 years ago.

• • •

THE PARTNERSHIP between humans and dogs has occupied anthropologists and archaeozoologists since their disciplines began. But such is the chronological uncertainty that clouds the debate, and so varied and intriguing the theories of when wolf part-bred into dog—when *Canis lupus* transformed into *Canis lupus familiaris*—that it is still open season. A new interpretation of events appears in the science journals every few months, each more assertive and convincing than the last, each with its own time line, but none of them wholly conclusive. But they're narrowing it down, sort of: it is believed that humans first began to domesticate dogs somewhere between fifteen thousand and forty thousand years ago.

The early visual evidence on rocks is both dramatic and naively beautiful, as most cave art tends to be, but it does not tell us when or why dogs evolved from wolves. And it certainly doesn't help us determine whether all dogs—all the hundreds of breeds in all their diverse physical makeup and with all their different roles and characteristics—all sprung from the DNA of one type of dog (perhaps the Canaan), or many. With dogs, as with much in life, art can only get you so far.

In mid-2019, a short list of the latest theories and discoveries—what happened when and how—reads like the middle of an unsatisfying detective novel, a police procedural, perhaps, in which the writer is edging closer to a rewarding and possibly devastating conclusion, but ultimately will leave the reader to decide for themselves from several possible endings. Some of the evidence is necessarily contradictory; it is the desire to eradicate these conflicts that drives the science on. We need to hold in our mind one thing at this point, which may be occasionally difficult for those of us who spend a lot of time in parks asking dogs, "And what breed are *you*?" We should remember that most dogs are composites.

In the estimation of Kathryn Lord, a postdoctoral fellow specializing in the evolution of dog behavior at the Broad Institute of MIT and Harvard, there are between 700 million and 1 billion dogs in the world. The great majority of these, probably more than 80 percent, breed of their own volition, without human interference, and most live near humans but not in their homes, scavenging off Dumpsters and trash bins. (It's an intriguing thought, albeit mostly a linguistic teaser, that most domesticated dogs are still living wild.) In warm climates the village dog would be short-haired and weigh about thirty pounds, while dogs in colder climates tend to be larger with longer and thicker coats. They are commonly a dirty yellow in color, with a white underbelly. They are much like the original dog, living in a similar environment to their ancestors many thousands of years ago.

Everyone now seems to agree that dogs have come to us from the gray wolf. The evidence for this has amassed gradually over the last century—a buildup of increasingly convincing opinion from a wide range of scientific disciplines—and it was confirmed by genomic analysis; to postulate another theory now would be akin to denying the existence of the moon. But leading scientists of the nineteenth century thought differently. In 1895, Nathaniel Southgate Shaler, dean of the Lawrence Scientific School at Harvard, wrote that "some students of the problem have inclined to the opinion that the dog is a descendent of the wolf; the whelps of this species, it is supposed, were captured by primitive men and brought under domestication." But Shaler disagreed.

> The difficulty of this view is that even with the high measure of care which the conditions of civilization permit us to devote to the effort, it has been found impossible to educate captive wolves to the point where they show any affection for their masters, or are in the least degree useful in the arts of the household or the occupations of the chase. They are, in fact, indomitably fierce and utterly self-regarding. It seems unreasonable to believe that any savage would have found either pleasure or profit from an effort to tame any of the known species of wolves.

Shaler also didn't have much regard for the popular Victorian theory that the dog was descended from a combination of wolf, jackal and coyote. Instead he believed that dogs were descended from a species he called *ancient* dogs. There were plenty of examples, he claimed vaguely, of species completely dying out before somehow reappearing as a different but more advanced type, but his evidence relied wholly on the discovery of dog skeletons a few thousand years old.

But what would be the *reasons* for domestication? Science still has difficulty in proving motivation. Nathaniel Shaler expressed the common view 125 years ago that humans initially welcomed dogs into their lives for companionship rather than usefulness, although present-day thinking reverses this. He believed that the first dogs, whose own motives were inevitably based on the availability of food around human camps, often became the focus of hunger themselves, being eaten when other sources of food became scarce.*

The current focus of research is on location and timing. "Maybe the reason there hasn't yet been a consensus about where dogs were domesticated is because everyone has been a little bit right," suggests Professor Greger Larson, the director of the palaeogenomics program at the School of Archaeology at Oxford. "Most animals were domesticated on a single occasion from a single wild population. What we have now is what we believe to be the first evidence, both genetically and archaeologically, that dogs were in fact domesticated two times." Larson's use of the phrase "two times" suggests a significance greater than "twice"; indeed, when his and his colleagues' work was published in *Science* in 2016, it caused more than a ripple of excitement.

DNA sequencing suggested there was a deep split between the genetic makeup of East Asian and Western Eurasian dogs. The genetic results were then compared to the archaeological record, and Larson found that there were

---

* Shaler was closer to the mark on another issue, though. Switching his thoughts from early domestication to the current position of dogs in society, in 1895 he concluded that "the mental qualities of our highly domesticated dogs are singularly like those of their masters, the likeness going to the point that the household pet is apt to have acquired something of the general character of the people with whom he dwells."

"very old dogs in the east, and very old dogs in the west, but in the middle it takes about 4,000 or 5,000 years after we first see them on either side of the old world for them to appear." The suggestion was clear: dogs were domesticated independently on two separate occasions from two distinct wolf populations thousands of years apart.

But this was far from the end of the story, for there was conflicting evidence that appeared equally compelling. At the end of 2015, just a few months before the Oxford team's report, researchers from Yunnan University in China published their results in the journal *Cell Research.* Using genome sequences from twelve gray wolves, twenty-seven primitive dogs from Asia and Africa and a collection of nineteen breeds from across the world, they found a far higher genetic diversity in dogs from southern East Asia than in other populations, leading them to declare that domestic dogs originated in southern East Asia 33,000 years ago. They believed that a subset of these ancestral dogs started migrating to the Middle East and Africa about 15,000 years ago, arriving in Europe about 5,000 years later. Also in 2015, *Proceedings of the National Academy of Sciences of the United States of America* (*PNAS*) published a report finding strong evidence that dogs were domesticated in Central Asia, perhaps near present-day Nepal and Mongolia.

Another study, published in 2017, and relying for its new findings on dog bones from two caves in Germany, again appeared to turn the work of Greger Larson and his Oxford team on its head. The first bone, from an early Neolithic site at Herxheim in the southwest, dated back

7,000 years, while the second, from late Neolithic remains at Cherry Tree Cave in the Bavarian uplands, was about 4,700 years old. The genetic pattern extracted from these samples—which were crossmatched with the DNA from a 5,000-year-old dog skull from a tomb at Newgrange, in Ireland, and almost 6,000 samples from modern dogs—suggested to Krishna Veeramah and his colleagues at Stony Brook University in New York that the dogs we know today all have a common origin, emerging from a single domestication of gray wolves. Moreover, an analysis of canine mutations over time enabled the researchers to genetically pinpoint a more precise time frame—somewhere between 36,000 and 41,500 years—probably from a wild, ancient type who were long extinct.

Dr. Veeramah agreed with the most accepted explanation for this divergence, that dogs evolved from the wolves who were scavenging on the edge of human camps when we were still hunter-gatherers. The less aggressive and tamer ones would have slowly been made welcome, and (in Darwinian fashion) have obtained more food. In time, a closer relationship with humans would become mutually beneficial: the dog-wolf welcomed not only a reliable food source but also the humans' desire to protect and nurture young dog-wolves, while humans would flourish from dog protection, transport (in the form of sledding), herding and hunting. In this way, humans and dogs evolved side by side. "We chose them, to be sure," the dog scholar Mark Derr suggests, "but they chose us too, and our shared characteristics may well account for our seemingly unshakable mutual intimacy."

The psychologist Alexandra Horowitz has found an even neater summation. The earliest dog-wolves "exploited a new ecological niche: us." Early humans ceased to be nomadic, settled in permanent spaces and threw things out. Horowitz notes that humans created trash right outside their settlements, and it wasn't long before the smartest dogs learned a little about the manipulation of humans. It's been going on ever since: you give dogs food, dogs will sit, roll over, whatever you want.

The last twenty years have seen many attempts to place this social interpretation in a scientific framework, and the science suggests that dogs were as much a product of self-domestication as they were of human engineering. An experiment conducted at Harvard in 2002 showed that dogs were far better than wolves at reading human communicative signals indicating the location of hidden food. Other recent experiments suggest that dogs and humans will share beneficial communicative eye contact from the puppy stage (when dogs face a task that is difficult to solve, they may gaze at humans to request help), whereas even after intensive domestic training wolves will struggle to offer a gaze that compares or compels.

Mark Derr has located another interesting and perhaps inevitable phase in this relationship—a formal divorce from the wolf population, a far more premeditated act than the one that bonded us. Once dogs and agriculture were firmly established, "the wolf became a competitor—an enemy, even—not because it was hunting us, but because it was taking our livestock," Derr suggests. "More recently, the conservation movement established a sharp divide between the wild and the built, a divide that really shouldn't exist, but does. At that point, the wolf became one thing and the dog became another, and they are in opposition rather than what they are, which is very closely related."*

In terms of evolution, humans prize anything that does not pose an immediate threat. In the case of dogs, scavengers were fine, but predators were not. With time, humans artificially selected and modified the scavengers into something more, namely a second domestication—the domestication of dogs in the home and the development of breeds.

A new piece of the jigsaw emerged in June 2019. A team of researchers, led by the cognitive psychologist Dr. Juliane Kaminski of the University of Portsmouth, had found an important difference in the facial muscle structure between dogs and wolves, something they believed had developed over thousands of years specifically to encourage and hasten domestication by improving communication with humans.

While the muscular anatomy of wolves and dogs was largely similar, a muscle responsible for raising the inner

---

* Among the less scientific interpretations, there is a mythical one from the Beng people of Ivory Coast. Near the beginning of the world, all animals once lived harmoniously together in a camp. But one day Dog found a rare and valuable egg among them, and he gathered it up furtively and hid it behind the Crescent Gate, the traditional venue for ritual objects. The egg hatched to bring forth man and woman, and the man soon made a gun, and he swiftly hunted meat. One hyena, fearful for his own life, hatched a secret plan with the other animals: they should destroy the Crescent Gate and everything within it. But Dog overheard Hyena, and told the man and woman, and so when the animals launched their offensive at the gate, man was ready for them and raised his gun. And that is why, from that day on, wild animals are all dispersed but Dog, man and woman are allies in a challenging world.

eyebrow was uniformly present in dogs but not in wolves. "Interestingly," the researchers reported in *Proceedings of the National Academy of Sciences*, "this movement increases pedomorphism and resembles an expression humans produce when sad, so its production in dogs may trigger a nurturing response." Pedomorphosis is defined as the retention of juvenile traits into later life: in other words, dogs had found a way to make the most of their puppy eyes, and look more like babies. The scientists hypothesized that the dogs' expressive eyebrows were the result of natural selection based on the preferences of humans. "When dogs make the movement," Dr. Kaminski noted, "it seems to elicit a strong desire in humans to look after them."

Earlier research on pedomorphosis had suggested that this arrested development lay at the very heart of why dogs gradually (over many centuries) began to look less like wolves. The most visible genetic imprint of wolves—pointy ears and a long snout—has now disappeared from most dogs, who have instead slowly emerged with snubbier noses and floppier ears. In 1997, Deborah Goodwin, John Bradshaw and Stephen Wickens reported in the journal *Animal Behaviour* that the more a dog resembled a wolf, the greater the tendency to display lupine behavior. Ten breeds of dog were analyzed for fifteen aggressive and submissive traits widely attributable to wolves. The researchers found that small breeds such as the Cavalier King Charles Spaniel and the Norfolk terrier, neither of which would ever be mistaken for a wolf, showed only two and three lupine behaviors respectively; the German shepherd, which was very much created to resemble a large wolf in appearance and aggressive tendencies, displayed eleven. There were one or two apparent anomalies in their findings—not least the golden retriever displaying twelve out of fifteen wolf-like behaviors—but this may be explained in terms of their ancestral forefathers: as hunting dogs, their old habits died hard.

And then we had another issue: as dogs slowly developed into human companions, it was helpful to distinguish one from another. Besides, a name bestowed a tag of utility—Ambush, Plotter, Guard—and, in time, a note of affection (Fluffy, Bella, Ludo). As we shall see in the following chapter, the names we call our dogs tell us much about the changing roles they play in our lives. Increasingly we will find that we are giving them the same names we are calling our children.

3.

Fido Thinks Maybe

In the twelve years that I've been walking with my dog, Ludo, on Hampstead Heath we've met:

Parker, Monty, Milo, Ronald, Alfie, Jenny, Wooster, Willow, Annie, Truffle, Randolph, Maxwell, Robbie, Lucy, Mango, Billy, Esme, Shnook, Toto, Menna, Kyffin, Tali, Rhian, another Billy, Boggin, Daisy, Honey, Snoop, Lucy, Oscar Brown, Calvin, Morley, Diggy, Pepper, Miles, Benjamin Barker, Jackson, Rothko, Margot, Colette, Pickle, Ellie, Kali, Penny, Big Arnie, Jessie, Little Arnie and—because this was Hampstead Heath and not Hackney Marshes—Swinburne. For much of the time that we knew him, Swinburne just sat under an oak tree and waited for the muse to come. Surprisingly, perhaps, Little Arnie is bigger than Big Arnie.

We've got to call them something, of course. There are approximately 9 million dogs in the U.K. and an estimated

90 million in the United States. My dog is also on good terms with dogs named Petra, Misty, Chia, Indian Chia, Truffle, Paolo, Herbert (Herbie), Rolo, Mischa, Leon Berger (the Leonberger), Gus, Moko, Pepper, Evangeline (Eva), Geoffrey, Pebble, Pobble, Slinky, Stinky, Honey, and Fog. To judge a dog by its name would be like judging a book by its cover, which one can regularly do with great accuracy. For example, a dog named Lord Rex will not be from the lower orders unless irony is involved. A dog named Stinky either spends too much or not enough time in the ponds. Dogs named Cupcake, Candy, Coco or Fudge will reflect an owner's sweet tooth and a rich connection between the comfort afforded by food and the comfort offered by dogs. And Fido, if he has any sense of self-worth at all, will rarely come back to you on first calling.

In this way we imprint upon our pets our human experiences and expectations. How and why do we apportion out these names? From whence Slinky and Misty? In all the following explanations, it is worth remembering that it is rarely the dog's fault. If the owner is keen on role-playing, then puppies who began life as Kate or Buster will wake up one morning as Brienne of Tarth. If the owner is a member of the London Library, then their dogs may too often be named a) Aurelius, b) Aurelia and c) Beowulf. Racing greyhounds appear to suffer the most, for an exceptional array of absurdity afflicts the super-fleet: past winners of the Cesarewitch greyhound race were called Future Cutlet and Jesmond Cutlet, and the parents of Jesmond Cutlet were called Lady Eleanor and Beef Cutlet. In 2009, the race was won by He Went Whoosh.

The key, one would have thought, is to choose a name one can call without embarrassment. A simple postmodern name like Spot or Bess or Pluto. Or choose a name that means something in an ancient or different language—Aurora, the Roman goddess of the dawn, perhaps, or Amaya, "night rain" in Japanese. These are grand public names, and their use will elicit a certain amount of ice-breaking with strangers. But with dogs as with humans, the public face will always conceal a private one, and the names owners use for their dogs in private reveal a propensity for baby talk and surrealism on a tremendous scale. For reasons I would not care to explain even if I could, my Labrador, Ludo, has been called (behind closed doors, with only select family members present) Human Zoo, Hoofus, Norkus, Lucy (odd for a male), Humphrey, Skelmersdale, Chairman and Herman.

Ludo was born in East Sussex in 2007, and he was part of an extensive litter. We were recommended to his breeders by a friend, but we were given no warning of the erratic sense of humor of his male breeder. After we had chosen Ludo from his squealing, nuzzling siblings, Mr. John Howe explained that because Ludo was only about a month old, he couldn't leave his mother for a few more weeks. To remember our choice, to distinguish him from the others, Mr. Howe said that he would cut off Ludo's left hind leg. We laughed. He repeated the joke as we drove away. We didn't laugh. Poor Ludo, hearing those words.

Ludo got his name after a family ballot and arguing. I wanted to call him Herbert, after my late father, which would then have been shortened to either Herb, Herbie or

Bert. I liked the idea of calling a dog Bert, although I'm not sure what my dad would have thought about it. My wife and the children in the household had other ideas, and I think in the end we whittled it down to a short list of three, and the two I favored most were rejected in favor of Ludo. The only two Ludos I knew were the hard-hitting current affairs journalist Ludovic Kennedy and Ludo the board game. I employed a pretty strong rearguard action by suggesting Cluedo, but that failed too. So Ludo he was, and Ludo he firmly became after about a week, and henceforth he could never possibly be anything else.

His full name—that is, his pedigree name sanctioned by the Kennel Club—is actually Greatcobwood Ulysses, poor chap. His parents were Willow of Parkdale and Greatcobwood Flora. His grandparents were named Autumnal Breeze of Glensue, Glensue Coney, Field Trial Champion (FTC) Dargdaffin Dynamo and Conneywarren Amber. His great-grandparents (as featured on his official five-generation pedigree certificate) were FTC Glenbriar Solo, Staindrop Scree of Glensue, FTC Broom-Tip of Carnochway and so on . . . The implicit aristocracy of these names is something we developed in Victorian times. It's a proof of breeding, of course, of both dog and owner, and I was delighted to see that Ludo came from a long list of show dogs—not because they had been shown, but that they had evidently been greatly cared for. The wonderfully absurd names are a mixture of their kennel name (Parkdale, Glensue, Carnochway) and the breeder's personal balance of soppiness, madness and decorum (Willow, Autumnal Breeze, Broom-Tip). One can only hope that, on the lam,

Staindrop Scree and Broom-Tip were affectionately shortened to Scree and Tip.*

In 2017, the most popular names for male dogs registered with pet insurance companies in the British Isles were Alfie, Charlie, Max and Oscar, while for females they were Poppy, Bella, Molly and Daisy. There was some crossover in New York, where, out of roughly 80,000 dogs registered in 2016, the most popular male names were Max (3,990), Rocky (2,769), Charlie (2,590) and Buddy (2,471), and the females were led by Bella (3,985), Lola (2,677), Lucy (2,379) and Daisy (2,240). Because it is New York, there were also 152 dogs named Biggie, presumably after the murdered rap star Biggie Smalls. (Presumably also, many of the dogs named Biggie were small.)

The French have a unique system for naming dogs—a reliable combination of simplicity and totalitarianism. Since 1926, the Société Centrale Canine has demanded that all pedigree dogs (*chien de race*) registered with them had to have a name beginning with the same initial according to the year of their birth. So in the first year there were an awful lot of Alphonses in the boulevards, and in the second year a great many Beatrices. In 2016 it was *M*, so the vets saw a lot of Madeleines and Marcels that year. In the clearest possible way this informed everyone—not least the dog, who was reminded of it a hundred times a

* All of which is to say, Ludo is far grander than sometimes he would like anyone to believe. But when, in the following pages, his owner promotes the benefits of owning a mutt, or the value of welcoming home a dog from a rescue shelter, I hope I am not being hypocritical: the intention is merely to present wider, hopeful options.

day—in which year they were born. In 2021 it will be *S*, so stand by for Sabine and Soleil (alas, you can't get your Fifi until 2030). There is one further madness: France has a terrible dearth of pedigree dogs named Klaxon, Yves or Zut, for in 1972 the letters *K*, *Q*, *W*, *X*, *Y*, and *Z* were retired on account of the difficulties people were having in finding enough names that *weren't* Klaxon, Yves or Zut.

THE NAMING of dogs stretches back to the ancients. Perhaps the earliest dog directory comes from Xenophon, the philosopher-historian born at the beginning of the Peloponnesian War in 431 B.C. Short names were best, he believed, making dogs easier to call in crowded forums and baths. He provided a long list of suggestions, although few have survived down the ages: Psyche, Thymus, Porpax, Styrax, Lonche, Lochos, Phrura, Phylax, Taxis, Xiphon, Phonax, Phlegon, Alce, Teuchon, Hyleus, Medas, Porthon, Sperchon, Orge, Breton and twenty-seven more, one of which was Speude. If you lost Phrura, Teuchon or Speude outside an amphitheater, how long would you be prepared to shout their names? And how long until your big yellow taxis turned up?

But Xenophon's names were not chosen at whim. They are descriptors, defining a particular temperament or ability. In the same order, the names above translate as Spirit, Courage, Shield-Hasp, Spear-Spike, Lance, Ambush, Guard, Keeper, Order, Darter, Barker, Fiery, Strength, Active, Search-Wood, Plotter, Ravager, Speed, Passion and Roarer. If ever proof were needed of a dog's usefulness, not least in the field of hunting or protection, here it is: apart from Passion, there's very little affection here, and no cuteness; Spear-Spike will almost certainly wreak havoc on the furniture, and you probably wouldn't want him too near your lap.

Some dogs have names that are so burdened with responsibility and superstition that one fears for their well-being, and for the very best of these we must call on the Beng tribespeople living in the Ivory Coast prefecture of M'Bahiakro. Here, we will not meet dogs called Max or Bella, but dogs called Kote Mo Nyré, Yreló, Bèkánti and Quelle Année. Kote Mo Nyré translates as "Problems don't seek people, people seek problems." Yreló means "Know yourself and avoid disputes."

The names hark back to the way ancient Egyptians regarded the dog as a deity, an animal who may foretell the most extreme and exacting turns of the world. Bèkánti means "If someone speaks ill of me I mustn't take it to heart." Quelle Année, which clearly derives from the French, and was a popular choice among the Beng, means more than "Which year?" It means "Which year will I become rich?" which is a difficult question for most humans to answer, and one would imagine a near impossible one for dogs.*

Dr. Alma Gottlieb, a professor of anthropology at the University of Illinois who lived with the Beng in West

* Readers of *The Far Side* by Gary Larson will be familiar with the concept of dogs wanting humans to acknowledge their inner lives. One cartoon is entitled "The Names We Give Dogs," and features a split panel. At the top a man explains to another man, "This is Rex, our new dog." In the panel below, three dogs explain that their names are Vexorg, Zornorph and Princess Sheewana. Sheewana's full name is Barker of Great Annoyance and Daughter of Queen La, Stainer of Persian Rugs.

Africa in the 1980s, observed that the tribespeople were rarely affectionate toward their dogs, despite holding them in high mythological regard; the dogs always had to fend for their own food. Kófla means "There are two roads for someone who is sick—maybe they'll get better and maybe they won't." And then there is the sad Beng favorite, È Tòé, which means—and you'd be brave to argue otherwise—"We're always swimming, but not all of us will float."

In the same belief system, Dr. Gottlieb notes, there are unique rituals regarding puppies. In the usual biological order of things, a puppy will open its eyes four days after birth. But if you're a puppy born in this part of Ivory Coast, you will only be able to open your eyes once a human being in the village has died and been treated as a divine sacrifice.

But the key point is surely this: whether walking on Hampstead Heath or the plains of West Africa, we keenly identify those things that are most precious to us. And upon those names we imprint our own interpretations of what a dog should be. The more human the name, or the more it embodies human attributes, the more respect we accord the dog in question, and the clearer our desire for our dogs to be like us. Spot is a dying breed; calling for Max and Bella in the park we could equally be calling for our dog or our child.

DO THE royals do things differently? Do our dog names take the lead from them the way our children's names seem to do? The answers are yes and no.

The Queen's Pembrokeshire Welsh corgis have become

a brand, as inseparable from the monarch as her handbags and pearls, and it is almost impossible to own a corgi in public without encountering an immediate reference to Buckingham Palace or Balmoral. The corgis have also become the object of mild and antiquated jest; there is nothing specifically funny to say about them, even though one feels there should be.

But the most notable thing about the Queen's corgis is that I have never met anyone who knows any of their names. There have been a great many corgis at the Palace over the years (the Queen alone has owned more than thirty during her reign), and many dorgis too, and yet hardly any of them have achieved individual fame, or even recognizability.* This is partly because a) these faithful little cattle dogs of Wales do look rather similar, and b) in the past there have been a lot of them scampering underfoot at the same time.

There have been two early exceptions. One was a dog named Dookie, born in 1933 but date of death unknown, the first corgi to have entered the household at the request of King George VI. Dookie was already royalty among the dog fancy: his mother was the champion Golden Girl, while his father was the champion Crymych President. But his

* The dorgi: a crossbreed popular with the royal family since the 1960s, when the Queen's corgi Tiny mated unexpectedly with Princess Margaret's dachshund Pipkin. Since then the family has seen the birth of dorgis named Tinker, Pickles, Chipper, Piper, Harris, Brandy, Berry, Cider, Candy and Vulcan. The dorgi, preceding the labradoodle by many years, may thus be seen as one of the earliest legitimizations of the unlikely crossbreed. The phrase "designer dog" is unlikely to have been in common usage at the Palace.

temperament was inclined to harmful truculence: he occasionally bit the legs of courtiers and guests, once drawing blood from Lord Lothian, former private secretary to Lloyd George, and he was equally fond of table legs. But Dookie didn't put the royals off corgis, because then there was Susan, the first dog Princess Elizabeth owned and cared for herself, possessor of a more even temperament, the mother of all who came after.*

One of the most eccentric official glimpses of the aristocracy and their dogs came in 1936 with the publication of the wonderfully obsequious book *Our Princesses and Their Dogs* by Michael Chance (with photography by "Studio Lisa"). The book is dedicated, on the suggestion of the Duchess of York, "to all children who love dogs," and appeared in the shops when the young princesses and their parents loved eight. One picture shows Princess Elizabeth on a bench with the corgis Jane and Dookie, both "very much alert as they watch a mysterious movement amongst the rhododendrons . . ." The following page places the same dogs on a lawn, "proudly conscious of the fact that the radiance of the Duchess of York and the sunny smiles of the princesses are a joy to everyone throughout the length and breadth of Britain." Another picture has Dookie, eyes closed, in some sort of rapture as Elizabeth puts her chin

* A dog resembling the modern-day Pembrokeshire Welsh corgi was first recorded at the beginning of the twelfth century, when Flemish weavers brought the dogs to Wales; another interpretation places the breed as a descendant of the Swedish Vallhund. Before they made intelligent, energetic and quite barky pets, they were reliable cattle herders. A recent increase in corgi registrations at the Kennel Club has been attributed to their appearance in the television series *The Crown*.

on his head. "Although inclined to occasional but harmless truculence . . . it would seem from this picture that he is also a born sentimentalist."

Early on, with the war imminent and then a reality, the dogs served as good PR for the royals. They normalized their humans; they hinted at responsibility; they enabled a show of emotion that the royal family had spent many generations repressing. But from the early 1950s, with Elizabeth on the throne and her immediate world relatively calm, the dogs became private, and when the public saw them it was incidentally, occasionally wandering across the foreground in a state photograph, sometimes accompanying a horsey and head-scarfed maneuver in the Highlands, but almost always anonymous. So here, with thanks to the faithful royal chronicler Penny Junor, is a list of names of the Queen's corgis in rough line of succession: Rozavel Lucky Strike, Rozavel Bailey, Honey, Sugar, Rozavel Beat the Band, Bee, Sherry, Whisky, Heather, Lees Maldwyn Lancelot, Buzz, Foxy, Convisat Endeavour, Mask, Rufus, Cindy, Brush, Kaytop Marshall, Rozavel Crown Princess, Geordie, Jolly, Sweep, Svottholme Red Ember of Lees, Smoky, Penmoel Such Fun of Rivona, Dime, Dawn, Dipper, Disco.

The names are both normal and bonkers, the champions distinguishable from the mere molting pets (and corgis shed *a lot*) by virtue of their pedigree title. We know not how these names were chosen or by whom; we can only wonder at the fun the name choosers had, and the names they rejected. Most remarkable of all, considering such edgy choices as Buzz, Geordie and Disco, is that they all descended from the Queen's very first breeding dog, Susan. But these names constitute less than half of the total. There is also Fay, Mint, Phoenix, Pundit, Socks, Plover, Wren, Larch, Laurel and Martin.* Most of them are royal corgis, a few are dorgis, and not all of them lived with the Queen. The line ended in April 2018 with the death of one called Willow from cancer at the age of fourteen, the Queen previously announcing her determination not to leave any behind after her own death.

But the Queen and her corgis were only a faithful continuation of the line, something the royals have always done well. Kingly associations with dogs go back at least as far as the hunting hounds of Henry VIII, but their domestic arrival in palaces and palace kennels began in earnest with Queen Victoria and Prince Albert in the 1840s, when noble-sounding greyhounds such as Eos and Laura, Swan and Helios, entered the frame. A Dandie Dinmont named Dandie, and dachshunds named Boy and Berghina, soon followed. Victoria's favorite dachshund was shipped over from Baden in 1872: his name was Waldman VI.

King Edward VII's terrier Caesar famously walked behind his coffin at his funeral procession in 1910, while larger aristocratic dogs such as Alex the borzoi, Sammy the poodle and Heather the collie accompanied the walkabouts of various British royals before the establishment of the house of Windsor in 1917. As well as their corgis and their young children, in the 1930s the Duke and Duchess of York also owned a Tibetan lion dog named Choo Choo,

* *Martin?*

and three yellow Labradors named Mimsy, Scrummy and Stiffy. Woe betide everyone within earshot when Stiffy went astray and had to be called home.

WHICH LEADS us to the story of Fido. You name a dog Fido and you burden him with as much social and moral responsibility as the corgis born to royalty or the mutts born to the Beng. But if you change a dog's name to Fido in its later years, then surely it must have truly earned that title. And so it was with a dog from the tiny village of Luco di Mugello, a few miles from Florence.

We first encounter him in a melancholy mood, stuck as he is in a roadside ditch. He is a young dog, probably no more than a year old, as yet unnamed, and he is in some pain, probably the result of a road accident. It is a winter evening in 1941, the middle of the war, and a brickmaker named Carlo Soriano is stepping off his regular bus from his nearby work. Soriano realizes the dog needs help as soon as he sees him, and so he takes him to the apartment where he lives with his wife, and together they do that beautiful human thing: they nurse an animal back to health.

That's one version. Others have given the man a slightly different name, Carlo Soria*ni*, and have placed the dog not in a ditch but under a bridge. The myth is already forming.

A few weeks pass. Back on his legs, albeit with a limp, the dog decides to dedicate his life to paying back his rescuer for his kindness. Each day after breakfast he accompanies Carlo to the bus stop from where he journeys to his job in Borgo San Lorenzo. Then the dog goes home. At the end of the day, as Carlo gets on the bus from work to take him back to Luco di Mugello, the dog leaves his home and ventures forth to meet him. The two then walk back together, and both of them enjoy their evening meal with Carlo's wife. This continues for about two years, and is utterly delightful for all who witness it. But then disaster strikes.

At the end of December 1943, Carlo's factory is hit in an aerial bombardment. He dies. The dog doesn't know that his owner has died, or at least is in denial of the fact. So each morning Fido—for so Carlo's wife and the teary villagers have now named him—leaves his home and ventures forth to meet his master. Carlo is not there. In the weeks, months and years that follow, Carlo is still not there, but Fido will not give up his vigil. He is a creature of habit and a creature of devotion, and soon he will be an international media star.

The reports began locally, and in April 1957, toward the end of his unusually long life, Fido was presented to the world by *Time* magazine. The report states that Fido didn't just wait patiently by the bus stop for his master to return each day, but under a bus, a habit which may explain how he got injured in the first place. According to the magazine, Fido had long been a popular hero: the butcher slipped him regular bones; when the night was bitter, a bus driver allowed Fido to accompany him on circular journeys and keep warm. Carlo's widow struggled to pay Fido's dog license each year, until, in the penultimate year of the dog's life, the mayor decided he should live tax-free, the only "legally unlicensed" dog in Italy. "He has set an example of fidelity to our village!" the mayor explained. As a tourist attraction, Fido would pay back the mayor many times over. He died on December 30, 1958, an improbable fifteen years to the day after his owner. A bronze statue now stands in the town square, and brings many romantics to a village they might otherwise have bypassed when motoring to Florence. Unbelievably, though, not everyone loves Fido; the statue was originally made of pottery, and one day some Campari-crazed hooligans smashed it up.

But by then Fido had become a movie star. In the 1950s a news crew filmed him walking to the bus stop, and what a beautiful thing it is to watch: the scrawny patchy dog, with a Labrador's muzzle and a spaniel's loose limbs, tail like a whip, limping his way through a courtyard, stirred no doubt by the rousing strings and commentary on the soundtrack, and then turning the corner into a heathland path and street. His daily journey is observed by another dog, who may be thinking, "Christ, not again, let him go Fido . . ." But our hero ends up at a street where a bus pulls up. Will Carlo be on it this time, a decade since he was blown up? We know "no." Everyone in the village, including all the other dogs, knows "no." But Fido thinks, "Maybe!!!" In a crueler and less romantic world, Fido would be diagnosed as a prime candidate for the madhouse.

The film—black-and-white, of course, and reassuringly grainy—shows Fido at the foot of a bus's door watching carefully as each passenger dismounts. It is heartbreaking to watch his perennial disappointment. No Carlo, no Carlo, no Carlo. No Carlo. The next scene shows a large crowd hugging and consoling Fido, several of them clearly distraught, one woman weeping into a dramatic white handkerchief, as if Carlo had just died that day, rather than ten years before. They are crying for their lost innocence. As the French writer Roger Grenier has noted, a dog may be a protection against life's insults, a defense against the world.

We need these Fido stories, both you and I. They make us both less and more alone. Comparable episodes of fidelity have come in from Japan, Scotland and the North Pole, and plump mayors will forever be queuing up to reward the most faithful dogs in their locality—a plaque here, a fountain there. In no small part we need the stories to confirm the loyalty of dogs, and their dependence on us. The stories are the ultimate display of unconditional love: Who wouldn't want to feel this sort of enduring devotion from a dog?

The behavioral traits that humans value most highly in dogs today may be summarized in a few rather unscientific words—"friendliness," "compatibility," "usefulness"—and one equally unscientific action, the disarming offering of a paw. These things, which may be further defined as loyalty and two-way nurturing, are the clearest expression of why we welcomed dogs into our lives to begin with. We first saw it in the extraordinary rock art in northwestern Saudi Arabia, and we've seen it in a million images since. And we see it so touchingly in the following chapter, where joyful companionship and scientific theory combine in the life and overshadowed work of the world's first great dog behaviorist.

4.

What Darwin Didn't Know About Dogs (Was Hardly Worth Knowing)

The family of Charles Darwin joked that he loved his dogs more than his cousins, and he never denied it. He always begged for news of his dogs when he was away from home, and although most of the time the updates were joyful, in February 1826 he received a letter from his elder sister containing news she feared to send.

Darwin was seventeen, unhappily enrolled in a medical course at Edinburgh, and he was missing his creature comforts, not least the comfort provided by his family dog Spark, whom Darwin called "dear little black nose." His sister Marianne was in charge of the terrier in his absence.

"The day after she came to us," she wrote to Charles, "she ran away, and though we made every possible enquiry and search, we heard nothing of her for a fortnight." Spark had taken up residence at someone else's home, and by the time she returned she had "become with pup."

"Last Monday the poor little thing was taken ill," Marianne continued, "and after the birth of one puppy she died. You cannot think how sorry we have all been about it. Everybody in the house had got so fond of her, and she was such a nice little dog." Spark's death had "vexed me more than any thing that has happened for a long time," Darwin's sister wrote, and she was doubly vexed about how Charles would take the news.

We do not have Darwin's reply, but it is clear from the subsequent family correspondence that he was downcast. Anyone who has lost a dog at a young age will know the feeling (losing a dog at a mature age is bad enough). But a dog for Darwin was not just a youthful passion. With the exception of the years he spent aboard ship, he owned or cared for dogs all his life, most of them terriers, but also an occasional Pomeranian or Scottish deerhound. Nina, Spark, Pincher, Shelah, Snow, Bran and Polly defined not only his need for loving companions, but his need for exemplars: more than any other animal—more than the creatures of Galápagos, or his barnacles and finches—dogs provided Darwin with inspiration and observational material (and it was a happy coincidence that the ship on which he sailed for five years was the HMS *Beagle*). Darwin was fascinated with how dogs existed in the world—their breeding, domestication and survival. But more than that, he was intrigued by how close to humans they appeared to be in their thoughts, expressions and psychological makeup.

Because of his other achievements, history does not tend to regard Darwin primarily as a canine behaviorist; he may have regarded himself as more of a canine hobbyist.

As with some of his other work, Darwin's observations laid him open to ridicule, or at least disquiet, but our present scientific understanding of how and why dogs behave as they do would do well to begin at his door. For often his intuition was spot-on. The bond between humans and dogs was evident in the attachment he felt toward dogs as companions, and the compatibility was confirmed by his observational experiments. Almost 150 years before the comparative unraveling of DNA samples from wolves and dogs made such a theory scientifically quantifiable, Darwin sensed a highly developed social aspect of a dog's makeup that didn't exist in their ancestors. Dogs caught Darwin's gaze the way they catch ours today: they are dependent and devotional, and we have little trouble interpreting their expressions—doleful, hopeful, excited—as something instantly familiar.

As he researched and wrote his third major work, *The Expression of the Emotions in Man and Animals* (1872), Darwin's terrier Polly slept in a basket by his feet. Sometimes she would wake to find that she had been included in his text; the references to "dog" in the index are based primarily on intimate notes jotted down in his study and on countryside walks. These include "the sympathetic movements of; turning around before lying down; scratching etc.; barking a means of expression; whining; drawing back the ears; gestures of affection; grinning; devotion, the expression of." Polly became a feature of his published work as much as of his leisure time; any dog owner who has written about canine behavior will know this pattern, but it is to Darwin's credit that so much of his behavioral analysis is still pertinent to all dogs we see around us.

Darwin's descriptions are infused with empathy. "When a dog approaches a strange dog or man in a savage or hostile frame of mind," he writes, "he walks upright and very stiffly . . . the tail is held erect and quite rigid; the hairs bristle, especially along the neck and back." But what happens when the dog realizes that the person approaching is his owner? "Let it be observed how completely and instantaneously his whole bearing is reversed . . . his tail, instead of being held stiff and upright, is lowered and wagged from side to side; his hair instantly becomes smooth . . . Not one of the above movements, so clearly expressive of affection, is of the least direct service to the animal."

His next observation, equally familiar and no less expressive, is the look of a disappointed dog when hopes of a walk are dashed by an owner with other plans. "Not far from my house a path branches off to the right," Darwin recalled, "leading to the hot-house, which I often used to visit for a few moments to look at my experimental plants." Until Darwin took this path, the dog was still hopeful of a walk, but as soon as he veered toward it, "the instantaneous and complete change of expression which came over him . . . was laughable. His look of rejection was known to every member of the family, and was called his *hot-house face*." The hot-house face involved a drooping head, a sinking body, falling ears and tail, a dulling of the eyes, and an all-round "piteous dejection."* Very few dog lovers will not recognize this pathetic and highly effective trait.

What may we learn from Darwin about, say, dogs that appear to be grinning? Today, Instagram and dog magazines are full of these images (golden retrievers and terriers in particular seem to be grinning all the time), and as humans it is impossible not be charmed by them. We know this is caused by nothing so much as the bone structure of their face, and they may be no happier than any other breed, yet the effect is infectious: for humans it is like looking in the mirror, the smile perhaps the ultimate signifier of the intimacy between us. Darwin reasoned that a dog's smile is rarely displayed "in a perfect manner," as it may be in a human, but is still easily discernible, the upper lip retracted to show teeth in a manner that might, if the action wasn't accompanied by tail wagging, otherwise suggest aggression. A connection—and sometimes confusion—between sniffing and sniggering was first noticed by the Scottish physician Charles Bell in *The Anatomy of Expression* (1844), and subsequently cited by Darwin in *The Expression of the Emotions in Man and Animals* (1872). "Dogs," Darwin writes, "in their expression of fondness . . . grin and sniff

---

* The prospect of anyone, especially Charles Darwin, laughing at their dog is unattractive. However, Darwin did propose—and he may have been the first esteemed scientist to do so—that dogs have a sense of humor. He distinguished it from "mere play" thus: "If a bit of stick or other such object be thrown to one, he will often carry it away for a short distance; and then squatting down with it on the ground close before him, will wait until his master comes quite close to take it away. The dog will then seize it and rush away in triumph, repeating the same manoeuvre, and evidently enjoying the practical joke." Not exactly a knee-slapper, but there we are.

amidst their gambols, in a way that resembles laughter." Of course, dogs have quite another way of showing laughter, grinning and happiness: they simply wag that tail.

And what about barking? "Under the expectation of any great pleasure, dogs bound and jump about in an extravagant manner, and bark for joy. The tendency to bark under this state of mind is inherited, or runs in the breed: greyhounds rarely bark, whilst the Spitz-dog barks so incessantly that he becomes a nuisance."* Darwin noted that a bark of anger and a bark of joy do not differ much, but may still be distinguished. He believed that with the exception of the *Canis latrans* species of North America (the coyote), barking was not an inherent behavior from birth but a learned one.

The Expression of the Emotions in Man and Animals was originally intended to form part of Darwin's previous book, *The Descent of Man* (1871). Here he expressed his belief that dogs could dream "a long succession of vivid and connected ideas," and that they possessed two psychological elements (the ability to regard inanimate objects as alive, and the willingness to submit to a higher power) that rendered them prime candidates as true followers of religion. His comparative ranking of dogs with the belief systems of "savages" necessarily plunged Darwin into hot water. There is still considerable pleasure to be gained from the description of his "full-grown and very sensible" dog encountering an umbrella. Lying flat on a lawn in a breeze, the umbrella

---

* The maligned spitz: probably of Arctic or Siberian descent, this pointy-eared, white-coated worker may often be found pulling sleds or herding.

appeared to be moving by itself. Had it been carried, Darwin reasoned, then his dog would have understood. But now he growled and barked, fearful and protective of his territory, and he must "have reasoned to himself in a rapid and unconscious manner, that movement without any apparent cause indicated the presence of some strange living agent."

Darwin goes further still, and his comparison between the religiously devout and the loyalty of dogs edges toward parody. "The feeling of religious devotion is a highly complex one," he writes, "consisting of love, complete submission to an exalted and mysterious superior, a strong sense of dependence, fear, reverence, gratitude . . ." He acknowledges that no being could experience so complex a range of emotions "until advanced in his intellectual and moral faculties to at least a moderately high level," yet there is clearly "a distant approach to this state of mind in the deep love of a dog for his master, associated with complete submission, some fear, and perhaps other feelings."

SMILING, BARKING, devotional submissiveness: modern science confirms the veracity of many of Darwin's instincts about dogs. But beyond that, it confirms the reasons for his personal emotional attachment toward his own dogs—we may as well call it soppiness—and it goes a long way to explaining his dogs' loyalty to him.

In 2017, the journal *Nature* referred to something called PDS—Pet-Directed Speech. This was very similar to IDS (Infant-Directed Speech), which parents use to talk to their children, and both were distinct from ADS, or Adult-Directed Speech. The journal published a report from four animal behaviorists in France who had previously wondered whether Pet-Directed Speech actually had any effect on the attentiveness of dogs—that is, does the particular human speech register incorporating a high-pitched voice and an increased pitch variation help dogs in their many tasks, or is it just annoying, both to dogs and to those who don't own dogs and believe that those who talk to their dogs that way are completely nuts?

The experiments involved filming forty-four adult dogs and nineteen puppies as they listened to a pre-recorded phrase spoken in PDS, IDS and ADS. The researchers observed how long their attention fixed on the loudspeaker relaying these phrases, something they called "total gaze duration." The results were clear, and maybe unsurprising. Although puppies seemed to respond to everything and nothing, adult dogs loved the higher frequency of Pet-Directed Speech, and it was suggested that it formed a quiet bargain with the dog: *talk to me that way and I'll do more of what you want.*

Darwin would have relished the researchers' conclusion: they found it likely that the effectiveness of PDS represented "an evolutionary determined adaptation" that benefits the regulation and maintenance of the relationships between owners and dogs. Darwin's son Francis would have recognized the findings in the behavior of his father. He recalled how excited Polly would get when reunited with her owner after a prolonged absence: she "would get wild with excitement, panting, squeaking, rushing around the room, and jumping on and off the chairs." And how did Charles Darwin react? "He used to stoop down, pressing her face to his, letting her lick him, and speaking to her with a peculiar, tender, caressing voice."

In their detailed report, the researchers discussed many other characteristics of the canine–human bond. Citing brain-imaging experiments conducted at Harvard and Azabu University in Japan in 2014 and 2015, they noted that both owners and dogs secrete oxytocin after a brief period of cuddling. They also detected a very similar brain activity when they compared mothers viewing photos of their children with mothers viewing photos of their dogs.

IN 1998, the French writer Roger Grenier observed that too close an attachment to humans may make domestic animals unhappy. "They spend their time observing their master, figuring out what he's going to do with them. Everything is a sign: a cough, a glance at a watch, turning off the television. There is no innocent act. Every minute carries its ration of anguish."*

How true this sounds. But now the tables have turned. These days, everything *dogs* do is a sign, monitored like never before. Every week a new dog theory leaps from the science journals. Dogs dream and it means something. Dogs beg for food and it means something more than "give me food." Dogs bark and it means a lot of things. In the media, a new theory of dog behavior is no less intriguing just because it remains unproven.

---

* From *The Difficulty of Being a Dog,* translated from the original *Les Larmes d'Ulysse.*

The popular dog psychologist Stanley Coren, for example, has identified five different types of bark, rooting them in ancient, protective, pack-like behavior. We will discover later how Charles Dickens became irritated to a murderous state when his writing was continually disturbed by barking dogs in the square below; anyone who has been sleep deprived by a neighbor's hound may share his bleak intentions.

Writing in *Psychology Today* in 2011, Coren defines the most common type of bark as rapid strings of two to four barks with pauses in between (this is the classic "alarm bark," meaning, *Call the pack—there is something going on that should be looked into*). Then there is a slightly slower bark in a fairly continuous string but with a lower pitch (a dog sensing an imminent problem, maybe an intruder or stranger getting closer; the dog is saying, *I don't think he is friendly. Get ready to defend yourself!*) Then there is barking quickly twice in a high or mid-range pitch (usually a welcome that means *Hello there!*, usually followed by the dog's "typical greeting ritual"; for my dog, Ludo, this usually entails an enthusiastic scampering around the kitchen in search of a toy he can present in his mouth to a visitor). Which just leaves a long string of solitary barks with deliberate pauses between each one (a sign of a lonely dog wanting company), and the stutter bark, which Stanley Coren suggested should sound like *Harr-ruff!*, and which is usually accompanied by the front legs of a dog flat out in front and the rear end held high; for dogs this is the best bark of all, because it means *Let's go play!*

But then again, dogs may bark for no reason at all. The ethologist Stephen Budiansky has written that "the amount of energy dogs expend in barking is phenomenal, totally out of proportion to any benefit they can possibly derive from the activity." This may be proof enough, he suggests, that barking may not be a purposefully selected learned behavior adapted through the ages, but something "that just happened to tumble out (along with floppy ears . . . and a certain amount of general goofiness) from the genetic mix-up that took place when wolves became dogs."

It may also follow that because a bark can signify a range of emotions, humans have no difficulty in interpreting barks as they see fit, naturally defining barking as the dog "talking." Even if a dog, especially a young one, barks for no apparent reason, they will soon learn that a bark may summon a sleeper, open a door, demand a walk. A whine may do the same thing, but a bark will probably get quicker results. The dog is swift to learn how consistently a bark may yield a result: bark at the postman for intruding on your patch and the postman quickly disappears; bark pretty much anywhere and it gets someone's attention. But Stephen Budiansky has noticed that dogs are "masterfully superstitious" at drawing connections between their actions and the results. "If you spend a lot of time barking, there are a lot of things that happen to happen while you're barking."

Sometimes it's the not-barking that counts. In 2007, the magazine *American Ethnologist* published a narrative report of Eduardo Kohn's stay with the Runa people in Ecuador's Upper Amazon. While he was there, the anthropologist woke one day to find that all three dogs of the family he was staying with had gone missing. A search in the forest yielded a terrible find: bite marks on the backs of their heads showed they had been killed by a jaguar. What puzzled his hosts was the fact that the dogs had not predicted their own fate by barking in their sleep. A woman named Amériga explained that her dogs usually dreamed each night by the fire, often barking *hua hua hua*. To her this would have indicated that they were happily dreaming of pursuing game to be brought back to the tribe. But if they had barked *cuai* that night, it would have been a sure signal that a jaguar would kill them the following day, for this was how they barked when attacked by big cats. But the dogs were silent before their death, and the tribe was concerned. The animals were an early-warning system, foretelling all sorts of things with their own language. Kohn reported that the Runa wondered whether, with their dogs absent, they would ever truly know anything again.

Here's another theory, and I'll claim this as my own: a dog barking for no apparent reason is the most annoying kind of bark to listen to, especially if it isn't your dog. Fortunately, there is plenty of instruction online for dealing with this, and some legislation. The website BarkingDogs.net has news items on what it calls the New York City "barking epidemic" of 2005, when more dogs than ever before caused headlines for refusing to shut up. At the end of that year, a new NYC law prohibited "unreasonable noise" from an animal for any extended time. Between 7 A.M. and 10 P.M. a dog was permitted to bark for no more than ten minutes, while between 10 P.M. and 7 A.M. only five minutes was allowed.

Fines of several hundred dollars were levied for contravention.*

BarkingDogs.net contains much no-nonsense analysis as to why your dog is barking, and the main message is clear: "Your dog is barking because you have placed him in a situation in which it is more rewarding for him to bark than it is for him to be quiet. In the final analysis, that is the only reason that any dog ever barks inappropriately." The site suggests that "your barking dog is not stupid," and his behavior makes perfect sense. The key is to change the situation to the point where your dog realizes that the most rewarding option is to remain silent.

There are several ways to quiet your dog, beyond moving him or shouting at him, the latter almost always counterproductive. There are collars that can automatically administer sudden shock-sprays of citronella or electricity. There are desperate operations in which surgeons may remove the vocal cords. And there is BarkingDogs.net's preferred method—"A gentle, little, corrective smack on the nose" or "some other humanely unpleasant thing [that] happens immediately after each and every bark. If you keep that up long enough, the dog will stop barking. It's that easy."

Perhaps, if you do not want an excessively vocal dog, and you do not wish to administer even a gentle smack on

* BarkingDogs.net includes "The Barking Dogs News Hall of Heroes," a short roll call of those who have done their best to silence loud dogs and their owners. (Among the honorees were three officials in Nahant, Massachusetts, who voted to ban two "extremely disruptive" golden retrievers from the town limits.)

the nose, some research into a dog's temperament should be undertaken before acquisition. The basenji, for example, although a handful in other ways, has been judged, in one notorious experiment, to be among the quietest of breeds.* And in the same test, the cocker spaniel came across the loudest: placed in a controlled environment with another pup, our hero barked 907 times in ten minutes, which is roughly one and a half yelps every second.

Why was this experiment notorious? Because, alongside many others, it formed part of the intensive period of dog scrutiny that took place over thirteen years in the 1950s and 1960s at the Jackson Laboratory at Bar Harbor, Maine. The work, which examined the development of dogs from birth, was the most detailed and prolonged study of genetic variations and canine behavior yet undertaken, and many of its findings still resonate. The most significant finding is that there is as much or more variation of behavior within a breed than there is between breeds.

But the experiments concentrated on only five breeds, so we must assume there will be far quieter dogs than basenjis, and far barkier ones than the cocker. And the researchers' initial period of breed selection was eccentric. They ruled out the smallest Chihuahua because of its low fertility and the largest Great Pyrenees because it would be too expensive to feed and maintain over long periods. Dachshunds and Scottish terriers were eliminated

* It's always swings and roundabouts with the basenji. A member of the sight hound family, the dogs are known for being peculiarly aloof, an attribute that humans have equated (so we're talking about feasible guesswork) with a high self-regard.

because they had a reputation for being stubborn. Dogs of roughly the same size were deemed most useful as they could use the same apparatus. The dogs they finally settled on were the basenji, the beagle, the cocker spaniel, the Shetland sheepdog and the wirehaired fox terrier.*

Among the experiments at Bar Harbor, the most enduring concerned the principles of early socialization. What was the ideal time for a puppy to leave its mother and bond with a potential owner? What makes some dogs fearful of human contact and others treat humans like their mothers? And what makes some dogs sociable toward other dogs and others seem perennially anguished? These questions have been asked since the sixteenth century, and they are of as much concern today as ever. The best answers usually derive from experienced breeders. But the results of the experiments conducted at Bar Harbor sixty years ago are still enlightening.†

Over the course of thirteen years, several hundred puppies were observed on a daily basis, the study lasting from

---

* These days, the Jackson Laboratory conducts its experiments on rodents. Its dog work culminated in 1965 with the publication of the classic (and still in print) *Genetics and the Social Behavior of the Dog* by J. P. Scott and J. L. Fuller.

† A few basic tenets have held firm through the centuries, such as those expressed in a manual by Wesley Mills, an American vet, in 1891: "In the training of puppies, first experiences are of much importance, and all the arrangements of the kennel, and in fact the whole environment, should be shaped in relation to this principle. The puppy should not be allowed to get into habits which will later need correction. Let him from the first be encouraged in cleanliness, self-respect, love of esteem, respect for the rights of other puppies, his fellows, etc."

birth to sixteen weeks; the researchers called it a "school for dogs," noting that it took five years for the school to be established sufficiently to produce eight years of peak performance. They saw that the first major change in behavior occurred at three weeks, when the sense organs became fully functional and the puppy began eating solid food (dogs are born both deaf and blind). As well as tests for socialization with humans (men and women) there were also experiments to gauge how the puppies socialized with one another and in larger groups, a bone being introduced each time to assess dominance and subordination. At five weeks, one puppy often emerges as the clear winner in a multi-pup tussle over the bone, and even more so at eleven weeks, but the dominance of particular puppies over each other changes between these two periods, and may continue to do so according to their size and motivational factors.

The more interesting test, known as the "wild dog" experiment, examined how puppies with early human contact differed from those who had hardly any for more than three months. The puppies were primarily reared outside, and came indoors for a week of human contact between the ages of two and nine weeks. The key period of initial socialization appeared to be around five weeks: the pups receiving human attention at that time displayed the least fear in the next few months, happily approaching their handlers and reacting well on the leash. Those brought in for a week of human contact between six and nine weeks also socialized well in subsequent weeks. As one might expect, the pups receiving barely any human contact for fourteen weeks were clearly fearful of it, and did not react kindly to being put on a leash; the researchers equated them with "wild animals."

These results would suggest that puppies are best housed with new owners at around six to eight weeks; one of the most common pieces of advice in popular dog-buying books appears to suggest that human bonding after two months may be detrimental to the relationship of both. But the crucial element is that dogs have at least *some* contact with humans in the first eight weeks (perhaps as little as thirty minutes twice a week in the first seven or eight weeks). There is no need for permanent housing with humans at this point. Indeed, there is evidence that separation from the mother at this stage may be detrimental to a puppy's self-confidence and general sociability; twelve weeks may be a far better time for this.

Throughout all this research—and this applies as much today as it did at the Jackson Laboratory in the 1950s—we should remember that dogs are dogs, as obvious as that sounds. There will never be a guaranteed consistency of behavior, just as we will never see this in humans. Environment and the temperament of the owner will always count as much as breeding. And we need to exercise caution when the performance of a dog is judged solely in human terms through human eyes.

AT THE end of 2017, researchers from the schools of psychology and life sciences at the University of Lincoln published their research into the similarities between the facial expression of emotions in humans and dogs, something Charles Darwin had attempted almost a century and a half earlier. A report in *Nature* explained their curious methods, which were distinctly old-school. In these days of fMRI brain imaging it surely took some chutzpah to compare a dog's facial reaction upon seeing the opening of a food container with a person's glee after winning a game show, but neither dog nor human seemed overly concerned.

The researchers set out to answer two questions: Do dogs display specific facial movements in response to different categories of emotional stimuli? And do dogs display similar facial movements to humans when reacting to emotionally comparable events?

And how does one begin to calibrate these things? In the mid–nineteenth century the French neurologist Guillaume Duchenne reported on what became known as the Duchenne smile (made up of genuinely heartfelt enjoyment), which differed by one muscle contraction from the non-Duchenne smile (less sincere, more formal). A smile could not therefore be classified simply as "happy," certainly not when attempting a cross-species comparison. A more recent universal scientific framework for measuring these things has existed since the late 1970s, and is still regarded as the gold standard. The Facial Action Coding System (FACS) has latterly been joined by DogFACS, and both use a category system for measuring the contraction or relaxation of individual facial muscles reacting to different events.*

---

* In the dog these muscles and fibers included the levator nasolabialis, the buccinator, the orbicularis oris and the caninus. DogFACS was coded with BORIS (Behavioral Observation Research Interactive Software).

The researchers in Lincoln measured five emotions—fear, frustration, positive anticipation, happiness and relaxation—in innovative ways. The triggers for humans included visualizing a dangerous animal (fear) and the gaining of a high monetary reward on the game show *The Cube* (happiness). The triggers for dogs included the visualization of food or hearing meal-related words (positive anticipation) and the initiation of a bout of play (happiness).

The facial reactions to these events were filmed and analyzed, and the results contained few surprises: dogs across multiple breeds displayed a wide range of similar and quantifiable expressions in response to certain emotional triggers. But were these expressions similar to those of humans? A lack of significant differences between humans and dogs would potentially be consistent with the shared basic origin of emotional expression as proposed by Darwin, or reflect convergent evolution. On the other hand, widespread significant differences would indicate that facial expressions of emotion are not consistent features of DNA across different species. The researchers found the latter to be true. When it came to dogs, Darwin wasn't right about everything: dogs produced distinct facial movements, but they were different from the ones expressed by humans in comparable emotional states.

Dogs possess a less complex facial musculature than humans, and we should be wary of the way we interpret and monitor their expressions. Fear and happiness are simple to gauge (from the tail as much as the face), but owners would be overhasty to believe they could easily detect guilt or embarrassment in the way they commonly like to do. It may well be that dogs do not "feel" guilt as much as express guilty looks as a learned behavior (hence drooping the head and moping away without eye contact).

In 2009, the psychologist and canine cognition specialist Alexandra Horowitz conducted an experiment to detect how and why dogs looked guilty. The process was simple: a dog was shown a treat by its owner, but was told not to touch it. When the owner left the room the treat was either eaten or removed, but when the owner returned they were sometimes told that the dog had eaten the treat when this wasn't the case, and they would thus rebuke their dog for something it hadn't done. The videotaped response from these rebukes showed dogs displaying classic signs of guilt even if they hadn't eaten the treat. "The effect of scolding was more pronounced when the dogs were obedient, not disobedient," Horowitz concluded in the journal *Behaviour Processes*. The so-called guilty look was "a response to owner cues, rather than . . . an appreciation of a misdeed." This may be another expression of domesticated loyalty and a dog's desire to please, and it certainly suggests another layer of cleverness: humans are getting the reaction a dog thinks they desire, and they will therefore not be ostracized from the pack, or park; instead, they may still get their dinner that evening. (Alexandra Horowitz also found that the dogs in her experiment who had attended a course of obedience classes displayed a far guiltier look in every circumstance than those who hadn't. They may be thinking: *After all that work, I'm still letting people down . . .*)

And after guilt, perhaps it was only fair to question

dogs about their guilty pleasures. In 2017, a group of researchers at the University of Glasgow published the results of their investigation into the effect of different genres of popular music on the stress levels of dogs in kennels. Their report acknowledged the previous work in the field, some of which they had conducted themselves, showing that classical music had indeed reduced stress in kenneled dogs, but it was short-lived; some dogs were only mellow for a day, and by day seven they were all their normal, anxious selves, Mozart or no Mozart. The purpose of the new study was to achieve and maintain lower levels of stress in the dogs, deemed important not only for their physical and mental well-being, but also in the display of the sort of desirable behaviors that could help facilitate adoption. So how would reggae, pop, Motown and soft rock compare to classical?* What sounds would relax the caged dogs more, and what sounds would get them dancing? Would the variety of aural enrichment in itself be enough to chill out the mutts?

There were other questions too, such as "Had the academic grant committee been just a tad overgenerous this time?" Possibly, but if nothing else we should be grateful for the light the work shines on the lives of dogs in rescue shelters. And if you were a kenneled dog, wouldn't you be grateful to anyone trying to improve your welfare in any way?

The study, which lasted eleven days, involved thirty-eight

---

* "Soft rock" is defined as Genesis or similar, although perhaps the dogs regarded this as "prog rock." Results might have been different with Fleetwood Mac.

dogs at the Scottish SPCA Dunbartonshire and West of Scotland animal rescue and rehoming center. There were twenty-eight males and fourteen females; nine of them were strays, seventeen had been declared "unwanted" by their owners, eight had been sequestered from owners because of welfare concerns, three had been returned to the shelter after attempted adoption, and one was there for temporary refuge. There were fourteen Staffordshire bull terriers, ten mixed breeds, five border collies, four lurchers, one border terrier, one Jack Russell terrier, one German shepherd, one Rottweiler, and one Rottweiler and Akita crossbreed. The dogs' prior time in the shelter ranged from 1 to 420 days.

The process was precise. For each genre of music, a six-hour randomized playlist was generated on Spotify and delivered through wireless Bluetooth speakers between 10 A.M. and 4 P.M. Measurements in heart rate variability were taken every hour, while urine samples (for gauging cortisol levels, the body's main stress hormone) were gathered at the beginning, middle and end of the eleven-day period. Observed behavior was also important, in particular how long a dog spent standing up or lying down to the music, and how long barking.

The results were generally encouraging, suggesting potentially beneficial physiological and behavioral changes from the playing of all genres of music. Heart measurements suggested a reduction in stress levels, and dogs spent more time lying down when they had access to Spotify than when there was silence. The magnitude of beneficial change detectable with heart monitors was highest for soft rock and reggae, followed by pop and classical, and lowest for Motown. The interval between heartbeats was significantly higher for soft rock than it was for reggae. But overall the response to different genres was mixed, raising the possibility, according to the published report in the journal *Physiology and Behavior*, that "dogs may have individual musical preferences." The researchers also found that the range of musical styles led to a longer period of relaxed activity than that previously measured when playing classical alone. And the published conclusion raised the possibility of another line of comparative inquiry in the future: audiobooks.*

Practically, emotionally, musically: clearly humans understand the workings of dogs so much better than they did before Darwin set sail on the HMS *Beagle.* And increasingly, as the next chapter explores, dogs have become far better at understanding and undertaking the work of humans.

---

* Perhaps there was a reason the researchers didn't play the dogs any rap. In its rhythm and its lyrics, rap is just raw music for dogs: intricate, bullish and relevant, it's the ultimate crossbreed form. It's not about Snoop Dogg or all the other dog artists (Bow Wow, Nate Dogg, Pitbull, Phife Dawg), or about the phallic and domineering references to big dogs in the songs, and certainly not those horrid old "bitch" lines. It's about stance and attitude, about putting it out there. Forget your Motown; rap speaks to and of a dog like nothing else.

5.

Dogs Will Heal

One Sunday at the end of August 2017, two metal detectorists named Pete Cresswell and Andrew Boughton excitedly unearthed a large selection of bronze fragments and coins in a farmer's field in the Forest of Dean in Gloucestershire, England. Their find included a section of a frying pan, a bronze animal paw from the side of a storage box, and twenty small pieces of a four-foot-high sculpture that appeared to have been smashed deliberately, allowing a fugitive Roman to hide and then re-smelt the scraps at a future date. Cresswell and Boughton alerted a local archaeology team to their discovery, and the objects then made their way to the Bristol Museum for cleaning and expert appraisal. A rare coin and other clues enabled the specialists to date the objects between A.D. 318 and A.D. 450.

The hoard contained one other remarkable item: a

bronze sculpture of an ornately engraved dog (definitely a he, with his genitals intact). The dog looked like an early prototype of the dachshund, a low stretched body over stout legs. The figure measured 5.25 inches high and 8.5 inches long (13.4 cm. × 21.4 cm.). In July 2019, two years after its discovery, the dog came up for auction at an antiquities sale at Christie's in Mayfair.

The dog had an estimated value of between £30,000 and £50,000. It was the last item in the auction, preceded by the standard array of Etruscan bowls and Minoan ax heads. There were about fifty people in the sale room, among them Christie's antique specialist Claudio Corsi, who was taking bids from clients on the phone. Corsi had noted in the catalog—and you couldn't really miss it—that the bronze dog had his tongue out. He may just have been thirsty, but it was more likely that he was serving a deeper purpose, a talismanic role as a healer. The precise location of the find had remained a secret to deter a rush of digs, but the site suggested to Corsi that it may have been connected to the Iron Age temple of Nodens at Lydney Park, where seven dogs with a similar function were excavated in the 1930s.

Dogs have been depicted as healers throughout antiquity. The animal was a constant companion of Asclepius, Greek god of medicine; in Roman times dogs were depicted as a healing attribute of Mars. The Celtic god Nodens was frequently associated with dogs and healing; dogs aided recovery by licking the wound of an injured person. Two small holes in the bronze figure suggested that it might once have been attached to a larger statue. Claudio Corsi wondered whether this was Nodens himself, or one of the dog's "patients."

Bidding was over within a minute, and there was a smattering of applause as the gavel fell; the final price was £137,500. The purchaser was a private collector, and he was swiftly congratulated by the two delighted men who had dug up the bronze in Gloucestershire. The little licking dog was in a square Perspex box on a chest-high plinth, and if he was finally happy to have found a new companion after 1,700 years in the earth and two years in transit, he wasn't giving anything away.

YOU WANT to meet a healing dog that isn't bronze? Just look around: most well-trained dogs provide that service these days. You stroke a friendly dog and you feel an indefinable sense of comfort. The dog probably feels it too, one of life's simplest, most liberating and joyous transactions. A dog binds us to the world like nothing else: to a history of all the dogs that have gone before, to a wider community that welcomes their presence and demands our selfless attention. This too is their purpose, why they are here with us and us with them.

But if you want to meet a canine healing *specialist*, a four-pawed provider of pleasure and placebo, one rewarding port of call would be the Whittington Health NHS Trust, an ungainly building in northwest London. Hundreds of dogs are having fun on Hampstead Heath just down the road, but some clearly have a higher calling, and one of them is Bryn, a lovely border collie, black and white on his body and brown on his snout, scampering his way down the oncology wing, endorphin with a tail. Bryn is a member of TheraPaws dog unit, part of the Mayhew animal welfare organization that arranges more than seven hundred dog visits a year to health centers and care homes, and sometimes to regular commercial offices where the staff are particularly stressed with work.

When you apply to become a therapy dog you have to meet certain criteria that would instantly rule out lesser dogs. For instance, do you react positively to meeting strangers? Would you let a stranger touch your ears and tail without getting flummoxed? Do you take food gently from a person's hand? Do you recover quickly from loud noises, or do you refuse to come out from under a table? Do you bark only rarely and have a calm temperament, or do you lose your cool on a regular basis?

Because he is a border collie, and not a Labrador, Bryn is not inherently calm, and does not at first glance appear to be the ideal therapy dog, but he is watchful and alert, and always keen to be of service. He is clearly a force for good in the hospital, where almost all the staff seem to know his name, and while they have to maintain a sober and professional air among their patients, they often revert to soppy fawning and child-speak in Bryn's presence. Whenever he visits Accident and Emergency, the emergency medicine consultant Dr. Heidi Edmundson posts a new photo of Bryn on her Twitter account. "Yesterday was a very busy and challenging shift," she posted the day after my visit. "So a big thank you to all the staff for their hard work. I was, however, delighted to get a visit (and some love) from Bryn. #therapydogs"

Bryn lives with Professor Caroline Fertleman, a pediatrics specialist. Four years ago the professor and her husband brought him home from a dog rescue charity, and he was a handful. He peed a lot indoors and ate stones. Professor Fertleman told me he is "much better" now, but he still gets easily bored, and he will often surprise the grown-ups with his antics, like the time they found him with a cat in his mouth.

Partly for this reason, Bryn doesn't stay in any one hospital department for long. I went on the rounds with him on a Tuesday afternoon, but his usual visiting day was Wednesday, so a lot of the inpatients weren't expecting him. They were still delighted to see him, though. We began in pediatrics, where some of the parents said they were allergic to pet hair and didn't want him to get too close, which Professor Fertleman said was the excuse adults often use when they are afraid of dogs, but most of the children were happy to stroke him. He jumped up for a cuddle on two beds, something that many dogs would give anything to do at home.

The process by which a dog becomes a therapy dog is thorough. They are observed in a domestic and working environment; are checked for fleas, vaccinations and so on; and then the dog's owner is checked for a criminal record and the ability to socialize. If approved, a dog must be free for perhaps ninety minutes at a time and should be available for at least six months.

As we did the rounds, and Bryn brought a distraction from what can otherwise be a very long and boring day, a parent of an inpatient wanted to know whether he was a pure border collie or some sort of mix. Professor Fertleman said she and her husband had spent £75 on a DNA test because they weren't 100 percent sure either (it turned out he was border collie through and through). The professor also said that while the patients loved seeing Bryn, part of his role was indeed to provide much-needed support to the staff as well, and when we went into various small offices with doctors sitting in front of screens they were all relieved to have a break and greet him.

Therapy dogs also provide therapy for the owners of therapy dogs—the calming balm of doing something selfless in a venal world, a de-guiltification action like carbon offsetting. And therapy dogs work all over. As a writer, I've always liked the example of Gunner, the book therapy dog. Gunner is a boxer who goes into schools and bookshops in the U.K. to encourage children to read and feel more confident about themselves. Gunner sits around admiringly as kids engage in stories, either by reading themselves or by being read to by adults. Between chapters, and sometimes between paragraphs, they are permitted to pat Gunner's head for encouragement and support.

Therapy dogs are nothing new: even Dickens knew a therapy dog. In February 1869 he visited a place he called "A Small Star in the East," the newly opened East London Children's Hospital, and Dispensary for Women in Limehouse. Not so long ago, Limehouse was the seeming epitome of viciousness, which meant crime, violence and abject, near-naked poverty.* Dickens took a shine to the amazing work being done at the hospital after a cholera outbreak, and he particularly admired a character he saw "trotting among the beds, on familiar terms with all the patients . . . a comical mongrel dog, called Poodles."

Poodles was "quite a tonic in himself." He had been found scrawny and starving outside the hospital, and had been nursed back to health. When Dickens saw him, the dog was wearing a collar inscribed "Judge not Poodles by external appearances," and he was wagging his tail on a boy's pillow. The hospital was run by a couple called the Heckfords, to whom Dickens credits the full exploitation of Poodles as a comforter and healer. For a while it appears that Poodles may have had an even higher calling, when one day "I find him making the rounds of the beds, like a house surgeon, attended by another dog—a friend—who appears to trot about him in the character of his pupil dresser." Dickens describes being led by Poodles to the bedside of a girl recovering from an amputation to restrict cancer. "A difficult operation, Poodles intimates, wagging his tail on the counterpane, but perfectly successful. . . . The patient, patting Poodles, adds with a smile: 'The leg was so much trouble to me, that I'm glad it's gone.'" Poodles moves on to examine another girl, this one with an enlarged tongue. Poodles puts out his own tongue sympathetically as he examines her "very gravely and knowingly." Declares Dickens: "I never saw anything in doggery finer than the deportment of Poodles."

---

* The hospital site is now home to a Holiday Inn Express.

At the Whittington, I wonder whether there has ever been anything finer in doggery than Bryn, especially if you are going through something as daunting as chemotherapy. I went with Bryn to visit a woman called Eve who was having drugs pumped into her while she lay on a bed wearing a tightly padded cold-compress helmet she hoped would save her hair. Bryn hopped on the bed to say hello, and Eve found it a nice diversion to stroke his head and ask Professor Fertleman about him. We then went to another patient who had finished her treatment but was having a top-up injection, and she was delighted to have a dog hop up on her bed too.

In the corridor there were patient survey cards pinned to a board. All of them gave the hospital top marks, and many of the comments were similar. "Loved seeing Bryn today. Cheered me up and helped finish my biscuits." One said, "Excellent doctors and nurses and lovely therapy dog! Very soothing and distracting, particularly the stroking." Another requested, "More Bryn please. Love him."

EVEN THE very earliest dogs began as service animals, although initially they only offered two services—hunting and many types of barking. But today a dog's ability as a service dog appears to be unlimited. Indeed, being a gentle therapy dog like Bryn may be the easiest job of all when one considers all the other things that dogs can do these days, such as hear for the deaf, see for the blind, detect the living beneath collapsed buildings, soothe those in severe distress and sniff for explosives, drugs and cancer tumors.

What makes the dog so perfect for these tasks? The answer is as plain as the nose on their face. As well as the dog's 200-plus-million olfactory receptors (compared to the human's 5 million), the dog has an entirely novel sniffing system.* Humans inhale and exhale through the same two nostrils, but a dog may inhale through its nostrils and exhale through side vents, the small slits enabling a far faster rotation of odors. In effect, the inhaling can continue indefinitely in a continuous stream, like the airflow depicted in an ad for a Dyson vacuum cleaner. Each nostril may work independently—stereo rather than mono. With the help of slow-motion video, researchers have formulated the notion of "expired turbulent canine nostril air jets," a process whereby even the exhalation of smells enables a fuller inhalation of new odor molecules.

Compared to humans, dogs require a far less dense concentration of a scent to detect it from a far greater distance. Another olfactory advantage may be found in a second sensing system just above the roof of the mouth called the vomeronasal organ, which is designed to detect hormone molecules; this helps dogs identify other dogs, and it is the reason dogs may pick up on a person's mood (to suggest, perhaps, when we need comfort and attention, or when we need space) as well as changes in health. The notion of a

---

* It seems that no one can agree on a precise figure: upper estimates stand at 6 million in humans and 300 million in dogs.

dog living entirely in the present is swiftly dispelled by their ability to detect an olfactory chronology at the foot of a tree, a condensed history of smells telling them who has passed by and who has left a calling card. For dogs, a smell reveals history and time. A wag once suggested that to pull a dog away from a lamppost while enraptured in smell was akin to dragging a scholar away from a rare text at the British Museum.*

Our celebration of dogs as versatile service animals is not a new thing, and no one celebrated their accomplishments with more gusto than the Victorians. In December 1891 the death was announced of a Scotch collie familiar to thousands of railway passengers from a large silver tag hanging from his collar. "I am Help," it announced, "the railway dog of England, and travelling agent for the orphans of railwaymen who are killed on duty. My office is at 55 Colebrooke Row, London. Where subscriptions will be thankfully received and duly acknowledged."

Help was at hand on the night boat train from London Bridge to Newhaven, walking the aisles and platforms with his message and his owner, the guard John Climpson, between them raising more than £1,000 for the orphan fund. Help's dedication inspired others to put their dogs forward for similar duty at all London's major railway terminals. For who could refuse such a willing volunteer on urgent duty?

A three-legged railway dog named Jack achieved fame

* As suggested in the film *Dean Spanley* (2008). The vomeronasal organ, which is also known as Jacobson's organ, is not unique to dogs. Elephants and reptiles have one too.

on the London and Brighton line in the 1880s, spending much of his time in the driver's cab, regularly brightening the days of railway staff, who regarded him as reason enough to turn up for work (according to a contemporary report, Jack's amputated leg was the result of a mishap with a mail train at Norwood Junction). The newsletter also reports that Jack was only loyal to railwaymen in uniform; he once returned to the south London home of a guard, only to desert him when he changed into regular clothes. The Brighton line appeared to attract all sorts of other strays too, also volunteer collectors for good causes, but not all of them trustworthy. A dog called Bob, for example, was adept at attracting coins from passengers and keeping them in his mouth until depositing them upon the charity collection plate. But good intentions sometimes went astray, and Bob frequently exchanged his coins for biscuits at a local bakery.

SCIENTIFICALLY, HUMANS have only begun to explore the dog's extraordinary powers of perception very recently. Unlike the cruel experiments conducted by Pavlov, say, or the heartless safety testing of chemicals in laboratories, the exploration has largely been a positive and humane one, certainly for humans, but largely for dogs too (although we do, of course, make such judgments on our own terms).

We are now at a point where we are logically asking ourselves whether there is anything a dog cannot do—and isn't happily willing to do—in the pursuit of maintaining or restoring human health. One celebrated example from

2001 concerns a woman called Connie Standley and her two Bouviers des Flandres (rough coat, stocky, bearded face). In 2001, Standley and her dogs were driving back after a trip to the Grand Canyon. They stopped at a fast food restaurant near her home in Florida, but Standley was told that her companions, Alex and Nathaniel, would have to return to the car, even after she explained to the owner that they were service dogs. Standley obviously wasn't blind, so the owner stood firm even when she told him about her epilepsy. The last thing he would want in his restaurant, she said, was for her to have a seizure.

The dogs possessed a remarkable ability to detect Standley's fits up to thirty minutes before they started, giving her ample time to locate a safe environment. Even her closest human friends were unable to provide such a warning. "Before I had the dogs I was standing in my kitchen drying a glass out and I went down in a seizure and put a glass right through my hand," Standley told the *New York Times*. But now the dogs warned her by tugging on her clothes, barking and jumping on her if she ignored them. Not long after she was refused admission to the restaurant, Florida passed a new state law permitting trained seizure dogs into all public spaces, placing them on a par with seeing and hearing dogs.

At the time of the incident there was much conjecture whether dogs could in fact provide an early warning, and some of this cynicism remains. A report in the journal *Seizure* in 2003 was optimistic that dogs might indeed be useful in early detection, yet it reported on a then-recent study which found that out of the twenty-nine dogs belonging to epilepsy patients only three had actively alerted their owners before an episode. A study at the University of Rennes, published in March 2019, again confirmed only the possibility of dogs' role in early detection, but did reinforce the extraordinary talent of their nose: their detection of a breast or liver tumor through the odor of a patient's breath was regarded as a relatively easy task compared to their ability to detect a change in odor from a person prone to seizures (on account of a fit's many possible causes and manifestations). "These results open a large field of research on the odour signature of seizures," the French researchers stated in the journal *Scientific Reports*, suggesting too the possibility of replicating the canine olfactory armory with an artificial machine known as an "e-nose." The electronic nose has existed since the early 1980s, consisting of delivery and detection sensors linked to a computer database.

To date, the e-nose is a replacement—a more scientifically consistent and measurable one—for the human nose, employed principally in the production, storage and monitoring of products such as food and other items prone to rot or contamination. The sophistication of these artificial techniques improves each year. They are still a long way off from equaling the detective performance of the least sophisticated nose of the least sophisticated dog.

And then there's the growing body of research suggesting that simply living and working with a dog may reduce the risk of heart attacks and strokes, if for no other reason than that the demands of owning a dog—regular exercise, responsibility beyond self-interest—may offer a calming, healthy and socializing effect; a dog benefits from an oxygenated walk in the park, a human benefits from an oxygenated walk in the park. Dogs are increasingly employed at schools and universities to reduce stress among pupils and teachers at exam time. Middlesex University, for example, employs five Labradors as "canine teaching assistants." The Labs each have their own staff cards and security passes. ("You can literally feel stress levels reducing!" reports Fiona Suthers, head of clinical skills.)*

On another campus, the headquarters of Amazon in Seattle, as many as seven thousand dogs come to work each day to support their owners and boost the mood of dog-loving staff. It's the cheapest and most effective therapy the company provides, the ability to feel fur at your feet on a weekday. All well-behaved dogs are welcome at the vast headquarters, where they enjoy their own outside areas and snack stations on every floor. Say what you like about the online behemoth, they know a thing or two about cozy marketing. When you land on an error page online and get

---

* How does one locate the ideal dog for such responsibilities? Traditionally, certain breeds—Labrador retrievers, golden retrievers and cocker spaniels among them—have been regarded as ideal candidates for specialized training. But even after an extensive course and careful socializing, more than half of those selected are later judged incapable of entering full service. It's an expensive failure rate. But in March 2017, a report in *Nature* suggested there was a more scientific method of selecting the ideal dog for the task. More than forty fully awake and unrestrained dogs from an assistance training facility in Santa Rosa, California, were placed in a Sphinx-like position in a functional MRI scanner, and various regions of their brains—including the caudate area responsive to reward activity—were examined for their responsiveness to commands both from their owners/trainers and strangers. Researchers found that a high motivational response to the signals of both trained individuals and strangers did indeed make it possible to predict suitability for assistance work.

a picture of a collie or dachshund, you know the company is killing your local store, but perhaps you soften a little.

The first dog at Amazon was named Rufus. Jeff Bezos employed a couple called Susan and Eric Benson—he ran the warehouse as the company's first managing director, she ran the book recommendation pages (when Amazon was still primarily in the books business), and their little dog ran around wherever he wanted. When the company expanded, the Bensons ensured that the name Rufus was written into the landlords' contracts as a welcome employee, and when other staff members thought it would be pleasant to bring in their dogs too, they hit upon a Spartacus-style scheme to gain admission: henceforth all dogs at Amazon would temporarily be known as Rufus too, and so in they came one by one, their owners unwilling to leave them behind as they sailed away in the biggest mercantile ark in the world.

We may logically ask whether a dog destressing an oncology unit or roaming freely through a customer complaints department provides much benefit to the dog himself. I think it does—if only because of the attention, if only because it's so much better than being left home alone. In one sense all dogs are service dogs, even—or perhaps especially—if that service is the provision of love and comfort. It's a virtuous circle, for in return the dogs get vats of attention back, and perhaps even a sense of purpose. Just as the Beatles sang, in the end, the love you take is surely equal to the love you make.

• • •

THE CANINE service that interests me most, not least because my youngest son has type 1 diabetes, is the hypo dog. Hypo dogs can detect when their human companions are in trouble with dangerously low blood sugar levels, and can alert them when they are no longer able to sense the danger themselves.

Hypoglycemia is the most common side effect of insulin management practiced by diabetics. A critical reduction in blood glucose may cause serious neurological and cardiovascular damage, and even the fear of a "hypo" may severely limit a person's ability to function normally. As a patient ages, and their neural alert system becomes increasingly immune to the early-warning symptoms of a hypo, the risk of damage increases. There is much anecdotal evidence that dogs—even those who have not been properly trained for this duty—may show behavioral changes when a hypo is imminent or occurring, and alert a person if an attack takes place while they are driving or during sleep. This intuition is useful but not assured, so it is natural to wonder how one may train an animal to be even more responsive to this danger. The clues lie in the detection of a specific change in body odors, usually a change in the person's breath.

In 2015, the American journal *Diabetes Therapy* reported on a small research project involving six dogs ages one to ten who had received a minimum of six months' training as Diabetic Alert Dogs (DADs). The dogs—two Labrador retrievers, one flat-coated retriever, a Siberian husky mix, a

German shepherd and a spaniel mix, who were also known as Carlie, Isabella, Jake, Juniper, Nala and Roscoe—were selected from shelters on the strength of their sociability, adaptability and confidence, and, as this was to be a test chiefly concerned with smell, on the shape and sensitivity of their noses. Their training involved some of the elements common to dogs learning to assist the blind: they were expected not to be alarmed by sudden noises or by interaction with strangers, and they learned how to lie quietly under a table as their owner visited an office or restaurant. They learned thirty commands, from the basic *sit*, *stay* and *lie down* to the more specific demands of finding help and retrieving emergency glucose boosters or a phone. And for two weeks the dogs were exposed to the perspiration and breath samples of people with type 1 diabetes, the swabs and airbags taken when they were both in a normal and a hypoglycemic state.

The researchers then rewarded a dog for sitting straight when she detected a hypo sample in a vial (she'd get a food treat). For the second stage, the vial containing the hypo sample was placed in a steel can, and again the dog would be rewarded if she sat straight in front of it. The challenge was repeated with additional vials in additional cans, one with the hypo sample, and others with either blank or normoglycemia samples. Finally, the glass vials were placed on a person, and the dog was taught to poke her nose into the person containing the hypo sample. More food treats rewarded a correct alert.

That was the basic training. The six dogs were then tested in a controlled environment, each filmed as they entered a room with a semicircle of seven cans arranged before them: one had a hypo sample, two had normoglycemia samples, and four cans contained a blank gauze pad. The dogs were rewarded for the correct identification of the hypo from an automatic food dispenser controlled by the experimenter outside the room, thus eliminating any human bias or contamination. The trained DADs were each tested eight times, and the results were both unequivocal and reassuring: the expected random success rate of locating the hypo sample was 14 percent, but the diabetes dogs reported a hit rate of between 50 percent and 87 percent.

The findings confirmed similar, earlier experiments, and the value to type 1 diabetics of owning a properly trained alert dog seemed inestimable: it could save your life. What was needed was more awareness among diabetics of the advantages of owning a dog, and a lot more trained dogs. This, of course, required more funding, and a little charitable heart-tugging, a challenge familiar to anyone raising money to train dogs for those with disabilities in seeing or hearing. The measured manipulation of human emotion is everything, something dogs have done rather well for centuries.

Consider this story of a young man called Tom and his dog, Freida. "About eight months ago you gave Tom a very special dog," Tom's mother wrote to Can Do Canines, an organization that supplies assistance dogs in New Hope, Minnesota. Tom has lived with type 1 diabetes from the age of seven and was unable to tell when his blood sugar was dropping. He had many seizures and many trips to hospital emergency rooms. "Until Freida came to live with Tom, I was terrified at night," his mother wrote. "I didn't want to go to sleep and miss a single sound: What if his blood sugar dropped and he didn't wake up? He could go into a coma, and without help from someone he could die."

And then Freida arrived, and Tom hasn't had a seizure or an ambulance ride since. "In only eight months he's already awakened him several times to let him know his blood sugar was low. She knows she has important work to do for him."

The cost of training a service dog like Freida to carry out these tasks lies in the region of £20,000 to £30,000, which seems like money well spent.

What drives a service dog? What causes a dog to be of such singular purpose, adopting a role far beyond what is normally expected of it? The answers are: a mind-set of selfless discipline encouraged by rigorous training; the supreme talents of an inquisitive nose; a not inconsiderable desire to make their owners happy; and a desire from their owners to justify and validate their own, human roles in a community. Humans created the first and last trait; the others come naturally to dogs at birth. Combined so beautifully, a dog's design has seldom appeared more noble, and we should be wary of taking it for granted. But in one sense—in the unwavering sense of companionship they provide—all dogs heal, and heal themselves. The bond we have created over the last few thousand years has confirmed this attribute above all others, and the strengthening of our relationship would be impossible without it.

But this too needs to be taught to be effective. A new puppy is not equipped to face the world without adult human guidance, and as the next chapter suggests, the expectations we now place on our dogs may be higher than ever before.

6.

The Smartest Dogs on Earth and Beyond

If I were a dog, I'd like to be trained by Susan Close. She would teach sitting and heeling in no time, but after four one-hour sessions with her I would also become familiar with such practices as fizz-up and freeze, door work, off-leash dancing, the unlucky dip, ping-pong, the flirt pole and the five freedoms. Cumulatively these elements and about thirty others would help me become a joy to both my human and canine companions and an asset to society; they would set me well on my way, at the age of about sixteen weeks, to being able to accommodate much of what life will throw at me. Henceforth I would not be scared of other people or noise, and I would not be spooked by black garbage bags when walking in the dark. I would be habituated and socialized, and people would love me so much that they'd insist on rewarding my exemplary behavior with a vast array of treats. In short, Susan Close would make me fat.*

Susan Close is seventy, stocky and dressed for dog handling, which is to say, not in a fancy manner. She is not a whisperer; her demeanor tend toward blunt and profane. She launched the Dog Hub for Camden Council in 2008. Since then she has amassed a stack of diplomas in the psychology of dog behavior and advised thousands of dogs and their companions on how to achieve mutual respect and mutual rewards, while also negotiating the perennially annoying daily problems of unwelcome pulling on the leash and failing to return when off it. "For instance," she says, "how many dogs have you seen where the owner shouts 'come!' and the dog doesn't come, and the owner gets angrier and angrier and starts shouting, and then the dog starts thinking, 'I don't *think* so!'"

Susan Close works in a small square room beneath a public housing project a short walk from Euston train station. The windows in the room are protected by metal grilles. An exit at the back of the room leads to an outside area for dealing with dogs deemed "out of control," marshaled by high iron fences. This remedial work, correcting the behavior of dogs who were maltreated by their owners (through cruelty or ignorance), used to occupy almost all of Close's dog time. Close has been bitten twice in her

* The "five freedoms" is a list of basic minimum requirements deemed essential for the care of all dogs and other animals (developed by the U.K. Farm Animal Welfare Council in the 1970s). They are: freedom from hunger or thirst; freedom from discomfort; freedom from pain, injury or disease; freedom to express normal behavior; and freedom from fear and distress.

work, but she blames herself for a naive approach, not the dogs. After a few years she had what she calls her "lightbulb moment," when she realized the best course of action would be to educate owners before they and their dogs became a problem. The natural bond between humans and dogs did not always incorporate best behavior from either party. The key was risk reduction through early training, and the puppy classes were born.

One wall of the Dog Hub's training room is piled with plastic boxes containing leads, muzzles and toys, which are dispensed by Close for no charge, while another box contains a sample of what not to use: steel collars used to inflict pain on a dog's neck if they pull on the lead and a collar that emits electric shocks. "For some people the only thing they know is punishment," Close says. "We see so many dogs that are fearful and anxious."*

Over the years, Close has compiled a very long list of books and pamphlets on errant dog behavior and its correction. Some of it is academic and observational, but most of it is practical and instructive, and it encompasses both simple, everyday advice for the novice and peer-reviewed scientific studies. Her book list includes such titles as *Chill Out Fido: How to Calm Your Dog*, and *Feisty Fido: Help for the Leash-Reactive Dog* (not forgetting *When Fido Sees Red:*

* Thankfully we have come a fair way from the training methods available to Victorians, when the talk was mainly about "breaking in" a dog. "Unlike most other arts," a popular manual by Lieutenant Colonel W. N. Hutchinson explained in 1850, "dog-breaking does not require much experience." But the experience that was required was often brutal. "Some dogs require constant encouragement; some you must never beat; whilst, to gain the required ascendancy over others, the whip must be occasionally employed."

Aggressive Behaviour in the Domestic Dog). The list of authors beginning with *M* alone runs to more than eighty publications, including *Reactive Rover: An Owner's Guide to On Leash Dog Aggression* by Kimberly Moeller and *Psycho Dog: All Your Dog Problems Answered in One Easy-to-Follow Guide* by Janet Menzies.

Taped on the training room window, behind the grille, is a notice from the Royal Mail warning owners about Fido's and Rover's predilection for postal workers. Last year in the U.K. there were 2,600 attacks. Dogs feel threatened by the uniforms and carts, perhaps, or by the slightly intrusive nature of a postal worker's task, or the dogs are disappointed that the post so rarely brings anything for them. (The advice on the notice is sturdy: move your dog to a calming safe room, do not let it meet the postman or -woman under any circumstance, and give it food at the precise time of the mail delivery.)

"I would want my dog to know it can't have everything it wants," Close told me. At home she has three Labradors. "I would want my dog to be comfortable about being touched, I want my dogs to make good choices, I want the dogs to be treated as viable individuals, separate beings, not just the cute puppies you bought for £2,500 on Gumtree or whatever."

Close has an exercise she believes would give new owners a clear indication of what their new dog experiences: place your phone at a puppy's head level and record a video as you walk down the street. You watch it back and "It's bloody frightening, especially if you're a Chihuahua."

She says there is no such thing as an easy or difficult

dog in her class. She is keen to never make a value judgment on a puppy, and says they all respond similarly to basic training. But she will make a value judgment on an owner. "I find some of them quite vulgar. Someone will arrive with their poochon and seem to think they have this very expensive magical thing in front of them and they lose sight of the fact that it's a dog.* I would say, 'Get that bow out of its hair and we'll talk . . .'" Increasingly, she says, dogs are treated as a possession rather than a companion. She finds that more and more dogs look like a baby, and she is thinking of the popularity of pugs and French bulldogs, dogs with flattened faces and a wider skull that may often have breathing problems.

I wondered whether half of Close's classwork was concerned not with the dog but the owner, and she replied, "Oh, ninety percent." She may not thank me for the comparison, but at times Close reminded me of the once-famous English dog trainer Barbara Woodhouse. Woodhouse was tweedy in appearance and stern in manner, and her popular no-nonsense approach did much in the 1980s to make a nation's animals come to heel; her attitude—and inimitable refrains of "Sit!" and "Walkies!"—may have had more influence on how Margaret Thatcher treated her cabinet than that of any other individual. "There is no such thing as a difficult dog," Woodhouse opined, "only an inexperienced owner. . . . I do not believe that a dog can be cured

* A poochon is what happens when a bichon frise meets a miniature or toy poodle. The characteristics are an animal with fizzing energy, a low boredom threshold and an almost suffocating cuteness. The history of designer dogs will be discussed in the following chapter.

by a psychiatrist, but I think some owners could be helped by one."

Not so long ago, Close suggested, a prospective dog owner would do their research and talk to a breeder, or perhaps go to a rescue center, but she was dismayed that one was now able to choose and pay for a puppy online and have it delivered to the front door, especially in the United States. Clearly not everyone acquires a dog in this way. But the trend is upward, and the influence of Instagram is worrying: increasingly Close and her trainer friends find that their classes contain dogs selected for looks, convenience and fashionability.

And there's another problem. "They can be so expensive!" Close said. "Someone phoned me up and they were getting a labradoodle from Australia—five thousand pounds, it will cost them altogether. And labradoodles are commonplace now, so people are looking for something else. It's the way of the world—you have to have the latest and most expensive model. When I was growing up those values were only attached to washing machines."

Susan Close believes that our changing relationship with dogs has been affected by our changing relationship with each other. People are more selfish now than they were, she reasons, which means some people expect puppies to arrive fully house-trained. "Somebody called me up at two in the morning and said, 'I got this puppy, it was delivered yesterday, and it's just peed on the floor and I can't cope with it.'" She says that owners will leave their dogs alone all day while they're at work, and are then surprised when they end up neurotic. I remembered a line from John Grogan, the author of *Marley & Me*: "Such short little lives our pets have to spend with us, and they spend most of it waiting for us to come home each day."

Close says she will never turn a puppy away, no matter how precious the owner. She and her assistants teach a basic four-week course sanctioned by the Kennel Club, but she maintains that her objective isn't just a well-trained dog but a well-behaved dog, a dog that will make judgments based on a wider understanding of its role and purpose. "Commands" are not the thing these days; the modern approach relies on "cues." But during her class, her moderation occasionally overheats. "You want to make me mad?" she says. "Refer to your dog as a *fur baby* . . ."

SUSAN CLOSE would no doubt get on rather well with Temple Grandin. Professor Grandin is well known for a number of things: her autism and insightful writing about the condition; her empathetic and campaigning work in the field of animal welfare and slaughter; and several groundbreaking books. A *Horizon* episode on Grandin was entitled "The Woman Who Thinks Like a Cow," but her thinking about the bond between dogs and humans is equally cogent.

One thing Grandin shares with Close is a distrust of the concept of the alpha male. They are fairly sure it exists in humans (in her book *Animals in Translation*, Grandin wrote of being molested by a then-prominent male psychologist), but they are dubious about the way it is most commonly used with regard to dogs. In *Animals Make Us*

Human, cowritten with Catherine Johnson, Grandin cites the pioneering work of David Mech, whose decades-long studies of native wolves in Canada exploded several myths about the way they behave in the wild. He stressed their remarkable friendliness, both toward each other and to humans, and he observed how they do not possess the alpha male pack mentality with which they are usually credited. He found instead a regular family relationship, the parents responsible for the upbringing of their pups, devoid of the adult male having to fight to establish his dominance. Our mistaken assumption of an alpha male hierarchy may have come from our observance of wolves in captivity, which are essentially non-family groups put together by humans.

Temple Grandin wonders whether this doesn't provide the strongest clue as to how dogs should also live in a modern environment. When dogs first evolved from wolves, they were able to do so only because of the existence of humans. Should the relationship in a centrally heated home be that different from the one that established itself in a cave or preagricultural camp settlement? In the home, dog owners needn't fear that, unless they establish themselves as a pack leader, the dog will somehow come to dominate the home; instead, human adults should adopt the same sort of parental role they would wish to provide for their children. It is this, rather than a role of dominance, that will also establish an ideal environment for the most effective training. Treating dogs like children—isn't that what we've been trying to get away from? Not if the scenario involves creating a nurturing atmosphere designed primarily to benefit the dog.

Intriguingly, Mech's work is new in its detail, but not so novel in its findings. In 1944, Adolph Murie's book *The Wolves of Mount McKinley* also found that wolves live naturally in families rather than hierarchical packs. These findings, old and new, provide another link in our understanding of domestication. While dogs evolved to live with humans, could it also be that dogs realistically evolved to live with human *families*?

SOMEWHERE IN the back of every trainer's mind there are two dogs named Rico and Chaser. These are the poster dogs of canine cognitive psychology, cited in hundreds of academic papers concerned with how much an individual dog can know, and how much he or she can figure out. Every canine behaviorist has reason to be grateful to them. Every dog has reason to regard them with a mixture of admiration and deep loathing, so high have Rico and Chaser raised the bar, so average do other dogs appear in comparison.

To fully comprehend their impact, we need to remind ourselves of a world where the intelligence of a dog was usually judged on the ability to sit or play dead. It was a world—or so it seems now—predominantly of cheap tricks and anecdotes, and perhaps no harm in that, for to remind ourselves of them now is still rather lovely. The letter in *Country Life* in 1917, for example, relating how Eleanor Peel was out fishing with her Scottish terrier one day when it began to rain. The dog didn't like the rain, and when his owner suggested a walk after the fishing, he began to limp.

"That dog understands every word you say—he is only shamming!" Peel's friend said. And sure enough, as they walked home, the dog decided the limp had done its job, and he started walking normally again.

This was intelligence up to a point, unless, as is rather more likely, the dog wasn't shamming and his limp just went away. In the same issue, a correspondent named Margaret A. White explained how her fox terrier used to bring her letters from the mat each day as she lay in bed. For this he would be rewarded with buttered bread. One day, when there were no letters from the postman, and thus no prospect of a reward, the dog deliberately knocked over the wastepaper basket and brought his owner an envelope from the day before. Buttered bread for effort alone.

And then, thirty-five years later in the same magazine, there was a story that almost defied belief. Stella, another terrier, this time Irish, was fond of her family and fond of holidays. Indeed, she was so fond of both that when her family left her at home one day when they set off for their annual trip from Nottingham to the Ingoldmells, a village on the British coast near Skegness, Stella decided to follow them on her own. A complicated day of subterfuge and hijinks ensued. Having walked down a side street to the nearest train station, she boarded a train to Grantham. She then crossed the main line by bridge and walked along a long platform to another train waiting in a siding. "Here, many mistakes could have been made," remembered one of the youngest family members, D. N. Stafford, but Stella boarded the correct train to Skegness, and was grudgingly let through the ticket barrier when the inspector failed to locate an adult who had bought her a special pass. She then managed to track down her owners, and bounded up the road to them with such enthusiasm that she knocked D. N. Stafford to the ground.

Stella appeared to be both hugely intelligent and extraordinarily fortunate. But this wasn't science, it was wish fulfillment. The stories of Rico and Chaser were a little different. In 2003, a woman named Julia Fischer was reading a newspaper in a café in Leipzig when she came across an item about a special edition of the popular German television program *Wetten, dass . . ?*, a show in which contestants had to bet on inconceivable feats. The show featured Rico, a border collie with a remarkable ability to remember the names of his stuffed toys. His owner, Susanne Baus, explained that since Rico had injured his shoulder as a puppy, she and her family had substituted rigorous outdoor exercise with a training regime in their small apartment. She would place three items in different rooms, and Rico would be able to retrieve whichever one her owner requested. By the time Susanne and Rico appeared on television, with Rico placed inside a large circle with his toys scattered all around it, he could locate seventy named items on command.

Julia Fischer mentioned this remarkable dog to colleagues at the Max Planck Institute for Evolutionary Anthropology in Leipzig, and they began to draw up experiments of their own. They had no doubts about Rico's talents, and there was no question of trickery involved, but their main interest lay beyond the bounds of entertainment. To what extent could Rico's ability to identify the assigned labels for specific objects be compared to the early learning of a toddler? In other words, could the cognitive ability of a dog be closer to that of a young human than previously thought possible?

First the researchers had to rule out what is known as the "Clever Hans" effect. Hans was a horse who became famous in Germany at the beginning of the twentieth century for being able to calculate sums and tell the time. Hans could indeed perform these feats, but only when its handler knew the answers to the math too: both consciously and inadvertently he was giving Hans visual cues from his posture and other emphases. Henceforth, psychologists set up their tests in controlled conditions where the trainer or experimenter was out of sight.

In Rico's case, the now two hundred items he was familiar with were placed in one room while the owner waited in another. The experimenter instructed the owner to ask her dog to bring two randomly chosen items one after the other from the adjacent room. Astonishingly, he retrieved a total of thirty-seven out of forty items correctly, and also managed to differentiate between commands to place the item in a box or bring it to a specified person.

But that was only the beginning. Julia Fischer and her colleagues Juliane Kaminski and Josep Call were also interested in "fast-mapping," the idea that Rico could think for himself not just by memorizing the names of objects he was familiar with, but by exclusion learning—identifying one new toy in a pile of familiar ones just because he hadn't seen it before. Rico was given a novel item to locate on ten occasions—for example, a caribou toy he'd never seen before—and on seven of these he successfully deduced what it must be. An even sterner test came four weeks later. Rico had had no contact with the caribou or his other new toys in the interim, yet he was able to remember half of the names of the items and retrieve them from a collection of both old and new items. "Rico had therefore learned that a word his owner had spoken to him only once was the name for a toy that he had identified through exclusion," Julia Fischer concluded. It was a retrieval rate she deduced was "comparable to the performance of three-year-old toddlers."*

When Julia Fischer and her colleagues published their report on Rico in *Science* in June 2004 and made Rico a star, they found that when the news spread to the wider media, most people weren't interested in the science or methodology. They wanted to know how to train their own dogs, and how their own dogs would compare against Rico. The answer was "not particularly well." A lot of dogs could identify ten toys. Rico not only knew the names of about two hundred, but also specific command words telling him where to place them. But a man called John Pilley had reason to believe that Rico was nothing special.

Pilley, a retired professor of psychology at Wofford College in South Carolina, had recently acquired a border collie named Chaser, a gift from his wife, Sally, to stop him from getting bored. Chaser, who was predominantly

---

* Intriguingly, a brief manual written by Thomas Wesley Mills in 1891 entitled *The Training of Dog and Its Psychology* included a similar thought: "The puppy at one period is like a young infant, later like a two-year-old child, and at the best most dogs never get beyond the intelligence of a young child in most respects, though in some qualities the wisest man is far behind the dog."

white with black patches and flecks, was so named because of her fondness for chasing. After a few months she had a daily routine, and it mostly involved one thing. "Play is the major reinforcer of Chaser's learning," Pilley said in a short film promoting the official history of his dog (the words are accompanied by a scene of Pilley spraying Chaser with a garden hose). "Too often, dog owners use food only as a reinforcer for behaviors. We have found that play is infinitely greater than food: it's not as distracting, and dogs don't satiate on play."

Why is the border collie so great at this? Primarily because of their experience attending to human commands when herding sheep.* To watch a film of Chaser obeying the commands of her owner on a patch of grass is in itself a remarkable thing—the crawling, the walking backward, the ability to take a single step and then stay—but to watch her find a single named object from a vast array is more remarkable still. It takes sixteen large plastic bins to hold all of her toys, and even before any experiments get under way, an observer might remark that they never knew there were so many soft toys available: every animal seems to be represented, from aardvark to walrus and zebra, and many imaginary ones and other objects too—a unicorn, a sunflower and something unidentifiable except by Chaser and Pilley (Pilley calls the object "Professor"; Chaser doesn't call it anything, but she knows it when she sees it).†

---

* As well as Rico and Chaser, there is also Betsy, an Austrian border collie. Her record was 340 words, but she could also identify fifteen people by name.

† Further echoes of Darwin: in *The Descent of Man* (1871) he wrote of conducting an experiment with his terrier Polly. He would ask her, "Hi,

In one example of Chaser's prowess, captured on film by the Wofford College psychology department in early 2009, more than a year before her achievements went public, experimenters are seen picking out a random selection of fifty toys from a pile of several hundred. The next scene shows Pilley kneeling on the floor holding a list of names on a scrap of paper, with several toys scattered behind him (thus eliminating the possibility of visual cues). A wily Chaser is ready to perform. "Find Sunflower!" Pilley says in a croaky voice. Chaser finds Sunflower and shakes it. "Good girl!" Pilley says. "Shake Sunflower!" Chaser drops Sunflower. "Find Cactus!" Chaser finds Cactus, which looks like a pineapple. "There's Cactus! Put in tub!" And so on, through Worm ("Squeeze Worm!") and London Bridge, which looks like a giant bee, and Poppy, which is a felt Poppy, and Bamboozle, which is an orange dog-looking thing with patches on its snout. Then Bozo, Sweet Potato, Big Rope, Professor, Poison Frog, Spook, Dapper Duck and Goose. "Squeeze Goose!"

Chaser learned the names of 1,022 objects one by one over three years, and after that she carried on learning more; she not only identified individual items, but was able to divide them into three distinct groups. The feat is as much a triumph of human patience as it is animal intelligence. Dr. Pilley teamed up with Alliston Reid, a fellow

---

hi, where is it?" and "she at once takes it as a sign that something is to be hunted, and generally first looks quickly all around, and then rushes into the nearest thicket, to scent for any game, but finding nothing, she looks up into any neighbouring tree for a squirrel. Now do not these actions clearly show that she had in her mind a general idea or concept that some animal is to be discovered or hunted?"

psychology professor at Wofford, to publish their results in the journal *Behavioural Processes* in November 2010, and reporters descended from all over the world. Grateful visitors would leave clutching Chaser's paw print made with an ink pad. When asked where he thought the boundary for canine learning lay, John Pilley looked both optimistic and weary when he said, "We think we're just on the frontier." But he also dispensed some down-home wisdom of his own: "I would rather be in a barrel of bumble bees than with a pessimistic person. Chaser is just the opposite—she's always happy, and that makes me happy."

Pilley died in the summer of 2018 at the age of eighty-nine.* When Chaser died a year later at the age of fifteen, her legacy was assured. *Paris Match* called her "the smartest and most beautiful dog in the world." Brian Hare, an evolutionary anthropologist at Duke University, has said that Chaser wasn't just performing "stupid pet tricks where people have spent . . . hours trying to just train a dog to do the same thing over and over." Instead, he regarded her as the most important dog in the history of modern science, which was surely both a boast and a challenge. For the space age has given us at least one other clear contender for this crown.

---

* There is a touching report on Chaser's Facebook page from one of John Pilley's daughters. "Many of you have asked if Chaser realized that my father passed away. She was with him every day while he was in hospice, she knew he was not well. Just hours before he passed, she uncharacteristically planted herself directly in front of his bed, stared at him and gave one very sharp, loud bark, continuing to stare at him. It startled us all and we looked at each other in astonishment. It wasn't 'wake up,' it was goodbye, and it gave us goosebumps."

• • •

LAIKA NEVER came home, and was never meant to. In November 1957, the first dog to orbit the earth was a Soviet hero, a space hero and a canine hero in that order, but she didn't die a hero's death. In fact, her heroism had tragedy built in: Sputnik 2 stayed in space for more than five months, but its inadequate cooling system ensured that Laika lasted barely five hours. She fried in her capsule, which wasn't quite the painless end officially attributed to her at the conclusion of a successful mission. The true circumstances of her death were obscured for forty-five years, not least, one imagines, for shame.

Laika was the name that stuck, though she had many others. *Laika* means "barker" in Russian, but at various times she was also known as Little Curly, Little Beetle and Little Lemon. The American press called her Muttnik. She could just as easily have been called after her rocket: *sputnik* means "companion." Laika's training was rigorous, and fell not far short of the routines afforded the cosmonauts who followed her: she was attached to heart monitors, strapped into corsets, inserted with a catheter, placed into air-pressure and isolation chambers, and spun around in a simulation of zero gravity. She was eventually strapped into a capsule so narrow that she could barely move her head. The photographs bring to mind those taken of Pavlov's many stray dogs in the 1890s during his classical conditioning experiments: similarly harnessed and attached to measuring devices, they do not suggest a set of happy dogs.

At the time of launch the *New York Times* called Laika

"the shaggiest, lonesomest, saddest dog in all history," although shaggy was hardly right—her coat was smooth and her profile was sleek (it needed to be to fit all the contraptions and capsules). And there had been a great many sad and lonely space dogs before her. Like all the earliest participants of the Soviet space program, Laika was plucked as a stray from the streets of Moscow, selected for size (small) and temperament (compliant). Before her came forty-four other dogs, most of them female, all traveling in pairs, most to an altitude of one hundred kilometers. The experiments began in 1951, and most of their fates are logged as "recovered safely," although occasionally there is "cabin decompression, both dogs died" or "parachute failed, both dogs died." Their names are unheralded, in Russian or English: Tsygan, Ryzhik, Knopka, Mishka, Kozyavka, Pestraya. "Neputeviy" translates as Scamp, "Otvazhnaya" as Brave One, "Veterok" as Little Fart.

After Laika there were thirty-four more. True fame was reserved for Belka and Strelka, who orbited the earth for a day in Sputnik 5 in 1960, and crucially returned alive to a state reception, and a heroic afterlife on stamps, candy tins and lampshades. When Yuri Gagarin became the first person to orbit the earth in April 1961, he is reported as asking, "Am I the first human in space, or the last dog?"

The dogs were hailed as the most intelligent human aids to scientific progress, true pioneers of space, true Cold War comrades. We should remember that they were not willing participants in this endeavor. But we should also note the great impact they had as simple animals: several interviews with leading Soviet space scientists conducted decades after Laika and Gagarin made history reveal a deep affection for the dogs, and not a little remorse. Many of those who survived lived the rest of their lives as honored guests in the homes of those who once had thrust them into the heavens.

7.

How We Got to Jackshi-tzu

As you might hope, the story of designer dogs begins with the original labradoodle. This pleasingly stupid name has been in occasional use since the 1950s, but it only became popular after one particular Labrador dated one particular poodle in Victoria, southeastern Australia, in the mid-1980s. Then, in the 1990s, the labradoodle became *really* popular, and in the following decade more popular still. And now, some thirty years since it first bounded onto the scene in a raggedy blur of adorableness, there is probably a labradoodle in every park in the world. Unfortunately, the story does not end entirely well.

The labradoodle began as a singular project with a unique objective. That the project swiftly ran out of control one may put down to a combination of all-too-common human traits: venality, ambition, a quest for innovation. At no point can the dog, or the dogs that came after, be blamed for anything, with the possible exception of excessive cuteness and a desire to please their owners. But it is precisely these traits that have boosted their popularity, and led to some terrible practices in breeding and welfare. More broadly, the success of the labradoodle may have skewed for good a dog's traditional belief that a human will predominantly have its best interests at heart.

The era of the designer dog, a phrase as awful as its concept, has produced an array of breed choices that would have been unimaginable to Charles Darwin and the scientists of Bar Harbor. We may take the sort of snooty and refined approach initially adopted by the Kennel Club and choose to dismiss them as a fad, while hoping in vain that the fad will pass. But this would be like dismissing the internet.

Even the ancient and vigorous bond between dog and human, resolutely reinforced over thousands of years, is not immune to the forces of scientific possibility and human desire. Indeed, these forces may have ensured the bond's robust continuance. Not long ago the idea of the designer dog (or "hybrid dog" or "diva dog") was vaguely shameful, clouded by association with eugenics, visited with the same sort of disapprobation that now accompanies canine cloning; a Hollywood pursuit, in other words, for the super-vain and idle rich. Not anymore. Such a dog has become a product, and our digital life has made even the most preposterous creation available for next-week delivery. A sample list appears both comical and increasingly absurd: labradoodle, cockapoo, yorkiepoo, springador, cockador, lhasapoo, frug, jackshi-tzu, chorkie, pomimo, borkie, bolonoodle, pooton, maltipoo, maltichon, malteagle, chonzer and schnoodle. One meets some of these dogs in the park and they are irresistible, primarily because so many resemble stuffed toys.

The burgeoning supply only exists because of burgeoning demand. While there are many experienced breeders producing healthy and beautiful dogs of this nature, there is also the opposite, those churning out dogs conceived in cruelty and laden with disease. Buying online from an efficient-looking site with hundreds of puppies for sale, a buyer will be influenced by cost and aesthetics. Dogs are photographed with their heads tilted to one side to increase cuteness and helplessness. We are already at a point where we may regard the purchase of a dog—a fragile addition to a human life for a decade or more—on a par with the purchase of a theater ticket.

Crossbreeding involves the deliberate matching of one breed with another to produce a dog that ideally combines the best attributes of both. Occasionally a program may be designed to reduce the risk of disease spread by generations of traditional pedigree inbreeding, but any manipulation of a gene pool may swiftly create its own problems. If a Labrador and a poodle carry the heritable hip and eye diseases that are common to their breeds, then casual and random crossbreeding will only exacerbate things.

In one sense, crossbreeding is nothing new: we've been doing it by accident or design since we lived in caves (what is a dog if not the manipulation of a wolf?). In another sense, the Great Dane and the Chihuahua are both products of the crossbreed. (When Mark Twain addressed his audience on a speaking tour in 1895, he described a dog he knew called Jasper: "He wa'n't no common dog, he wa'n't no mongrel; he was a composite. A composite dog is a dog that is made up of all the valuable qualities that's in the dog breed—kind of a syndicate; and a mongrel is made up of all riffraff that's left over.")

Perhaps all that's changed is our taste, and the unregulated and accelerated speed of the breeding programs. But one other thing has changed too: until fairly recently, a dog's purpose was rarely cuteness alone. Dogs hunted, guarded, herded, tracked, guided, detected, restrained, fought and comforted, but the posing for photographs on Instagram is a relatively new development.

THE KENNEL Club has strict rules as to what constitutes a breed: it only added the Jack Russell terrier to its register in 2016. And because the organization declines to recognize the new designer crossbreeds as pedigree dogs (sometimes referring to them instead as "types"), and because it refuses to issue registration certificates documenting their lineage, the proud owner or breeder of a designer dog has felt obliged to look elsewhere for official recognition. The search does not take long; no paying customer goes away empty-handed.

Anyone requiring evidence of how fantastical things have become need only consult the perversely energetic options once provided by the International Designer Canine Registry (IDCR). This extensive list begins with the affenchon (the result of an affenpinscher crossed with a bichon frise) and ends—more than six hundred dogs later—with the ewoak (Yorkshire terrier × zuchon teddy bear).* There are thirty-two beagle permutations, ranging from the beagle and schnauzer combo (schneagle) and the beagle and shih-tzu edition (bea-tzu), to the beagle and miniature pinscher (meagle) to the beagle and French bulldog (frengle). Those who love the frequent appearances of the expressions "poo" and "oodle" will be delighted to find that the poodle has been officially blended fifty-four times, including those blends arising from very popular visits with the wire fox terrier (resulting in the wire foodle) and the Catahoula leopard dog (voilà, the pooahoula). Aficionados of toilet humor will only be delighted with the poodle-Havanese extravaganza known as the havapoo.

The poor basset hound has only been remixed thirteen times, but the list does include the almost impossible and absolutely disgusting coupling of the basset and the dachshund, resulting in the still-near-impossible (to pronounce, if nothing else) basschshund. You probably didn't want to be there when the basset and the Chinese Shar-Pei got it on, or the basset and the chow chow. In my traumatized mind I can hear a lot of human voices shouting encouragement and sprinkling petals on pillows, and it is difficult not to be bemused and horrified in equal measure (horrified because the reality, alas, is very far from this romantic notion).†

* The zuchon teddy bear is itself a cross between the bichon frise and the shih-tzu. Is there no end to it?

† The Kennel Club does now provide an alternative to this Wild West officialdom: while still denying a full pedigree certificate to these new dogs,

In addition to its certificates, the IDCR provided a brief guide to the temperament of all the dogs it certifies. So the cavamo (American Eskimo × Cavalier King Charles Spaniel, first registered in 2009) gets four out of five stars for Other Pet Compatibility, three out of five for Activity Level, Grooming Necessity, People Friendliness and Trainability, and two for Shedding Level and Noise Level (two being not very barky or whiney). But if you want a quiet life you certainly shouldn't get a jackapoo (poodle × Jack Russell terrier). This one gets five stars for Noise Level. But as it also gets five stars for Trainability, presumably one could train it not to bark so much. This is evidently not an exact science; the phrase "blind guesswork" may be nearer the mark. The IDCR is less active than it has been, but you will find many similar sites online with even more, and newer, breed options.

Designerbreedregistry.com offers certificates for the muggin (miniature pinscher × pug) and the giant bolonauzer (Bolognese × giant schnauzer) alongside a semi-inspirational, wholly emetic quote on its homepage: ". . . from those who dream . . . and are not held back . . . from those who have the passion and drive to create . . . to the trailblazers who have a vision . . . Creative instincts have given humanity a very diverse collection of Man's Best Friend to accompany us on our walk."

Fans of these preposterous creations may argue that the world should be filled with nothing else; if they're

---

they are entitled to a special registration to enable them to take part in Kennel Club–endorsed activity competitions.

well looked after in caring homes, then why not? Walking in the park, and in fashionable urban areas, you will encounter a great many of these new dogs, and they will charm the pants off you. They may even charm the pants off themselves, for a great many will be wearing clothes of some sort, perhaps a woolen body warmer, almost certainly a neckerchief, quite possibly an outfit bought online that will make a bolonoodle and malteagle look like a dinosaur (put "dinosaur dog coat" into Google Images and pick up your jaw from the floor). These dinosaur dogs will not die from extinction any time soon, but instead anticipate a long and prosperous life. The only thing that will kill them will be suffocation by mollycoddling.

Alas, the most mollycoddled dogs have their opposites. There are just too many stories about maltreatment to make cuteness an acceptable purpose. There have been terrible reports of puppy farms churning out dogs who are blind, deformed and disease-ridden, and of tiny sedated dogs trafficked through borders in near-suffocating conditions for profit alone. A single purchase endorses and extends this chain.

And then there are "teacup dogs," so small and so light (five pounds or less) that they may sit in a palm or be carried in their owners' handbags; they are often sold as "apartment dogs" with no need to venture outside, which makes them almost cats. They are bred from dogs that are naturally very small to begin with, but often from the runts of litters affected by birth defects. They will be susceptible to heart defects, liver dysfunction, hypoglycemia, respiratory problems, blindness and something known as patella luxation, which means that if these dogs do ever make the leap from handbag or other receptacle to pavement and park, they will be unable to walk. The question "What is wrong with these dogs?" is easier to answer than the question "What is wrong with these breeders and owners?"

THIS IS how the saga began. In the early 1980s, a man named Wally Conron received a request he thought he could handle in his sleep. He worked as the breeding manager for the Royal Guide Dog Association of Australia in Victoria. The request was to provide a vision-impaired woman in Hawaii with a guide dog, but also to ensure that her partner, who was allergic to pet hair, could live with the dog too.

"Piece of cake, I thought," Conron recalled more than twenty years later. "The standard poodle, a trainable working dog, was probably the most suitable breed, with its tightly curled coat. . . . It turned out I was wrong." Conron, who was then in his sixties and hugely experienced, spent two years testing various poodles, conducting more than thirty trials. But he still couldn't find one to reliably serve as a guide dog. He decided "in desperation" to cross a male poodle with one of his established Labrador retrievers. Three puppies resulted, the hair of one of which didn't make the Hawaiian woman's partner sneeze, and a new program commenced. The dog was named Sultan, and in time he would cause a revolution.

It was hoped that all three puppies would become guide

dogs, so first they had to be socialized for a year with families. Once again, Conron was wrong about how easy this would be. "It seemed no one wanted a crossbred puppy; everyone on the waiting list preferred to wait for a purebred." Also, no one had seen these types of dogs before; to many people they just looked wrong.

With the puppies still homeless after eight weeks, Conron had an idea. "I decided to stop mentioning the word 'crossbreed' and introduced the term 'labradoodle' instead." Things changed. "During the weeks that followed, our switchboard was inundated with calls from other guide-dog centers, vision-impaired people and people allergic to dog hair who wanted to know more about this 'wonder dog.' My three pups may have been mongrels at heart—but the furor did not abate."

Predictably, there was less interest from traditional pedigree breeders. Rather, disdain. When Conron sought to exploit his success by obtaining more poodles for an extended breeding program, the governing Kennel Control Council of Australia told him that any established breeder supplying dogs for his program would be struck off its register. When Conron assured the KCC that the resulting dogs would be bred not for profit but to help the blind, and that Sultan was now proving to be an invaluable aid in Hawaii, it made no difference.

Some breeders did see the value of Conron's program and offered their male poodles for his Labradors, but then there were other dilemmas: the next litter of ten labradoodles produced only three offering freedom from allergy. Over the years, as the program refined itself, demand continued to grow. It didn't do any harm (one would have thought) that labradoodles look so attractive in photographs. Conron initially enjoyed the attention, and he took satisfaction from his work. And he was having fun: he crossed a labradoodle with a labradoodle and called it a doubledoodle; he called the next generation a tripledoodle.

Soon enough, the inevitable happened: there was competition from what Conron called "backyard breeders." "Were breeders bothering to check their sires and bitches for hereditary faults," he pondered in the Australian *Reader's Digest* in 2007, "or were they simply caught up in delivering to hungry customers the next status symbol?"

WHAT DISTINGUISHES a crossbreed from what we once called a mongrel or a mutt? Intention. And semantics: it's often said that a dog from two pedigree breeds is a crossbreed, while three (i.e., with a mother produced by two breeds and with a father from a third) is a mutt. But this definition is flimsy and complex: a lurcher, for example, is not regarded as a pure breed, but rather a cross between a dog from the sight hound and scent hound group, or perhaps a terrier.*

There are over 337 dog breeds recognized by the Fédération Cynologique Internationale (FCI), the Belgian organization founded in 1911 to represent a host of kennel clubs

* An ancient and noble grouping, the sight hound clan includes the Irish wolfhound, the Afghan hound, the greyhound and the whippet. A scent hound is a traditional hunting dog such as a beagle, a bloodhound or a foxhound.

worldwide. The Kennel Club of Great Britain recognizes 212 pedigree breeds, while the American Kennel Club logs 192. Significantly, the names of many officially recognized traditional pure breeds are now more obscure in both name and quantity than those popularized by Wally Conron. We make fun of the portmanteau names of designer dogs, even though the joke is wearing thin. But we do not get the last laugh, because the cavachon, the cavapoo and the goldendoodle are already more recognizable (both in moniker and physical form) than the pedigree Schweizer Laufhund (a Swiss hunting dog, exceptional nose, resembles a taller, stretched beagle), the Treeing Walker Coonhound (a descendant of the foxhound; proficient at pursuing game, especially pursuing game into a tree until it can be extracted by a human; frequent barker, often with a smooth black/white speckled coat and a bit raggedy), the Lancashire heeler (small and squat, not unlike a corgi in a darker coat, upward pointed ears, good ratter trained initially to nip at heels as they drove cattle to market, now an affectionate if stubborn family dog), the Slovenský kopov (a beautiful scent hound, largely Slovakian, black and tan coat, muzzle like a thin Labrador, large ears, courageous hunter), and the porcelaine (white, shiny, short-haired coat with light black mottling and light brown ears, French/Swiss in origin, keen hunter, long proud neck).

And we should remember that we have been at this crazy juncture before, or very near it. In January 1899, an illustration in *Punch* parodied the variety of strange creatures that were now popular in London's royal parks. A fashionable woman is seen walking alongside eight dogs of varying size and temperament, all vaguely recognizable, but all also somewhat changed. There appears to be an Irish wolfhound, a Chihuahua, a dachshund and a bulldog, but the dachshund is scaly and the bulldog has tusks like a walrus. A caption reveals their true identity: "Dorgupine, Crocadachshund, Pomme-de-Terrier, Ventre-à-Terrier, Hippopotamian Bull-dog, German Sausage Dog, Hedge-Dog, Bug-Dog."

The extremes of today, enabled by technology and driven by the pursuit of the new and unique, are just that: extremes. *Punch* exposed the desire for novelty at the end of the nineteenth century, but the trend was already facing raised eyebrows at the beginning of it, on at least one occasion by dogs themselves. *The Dog of Knowledge*, published in 1801, purportedly written by a terrier named Bob, contained a scathing passage in its first chapter describing how the author came to be.

The terrier is an ancient breed, and hardiness its main characteristic; their wiry fur was usually a weathered tan color. But things have changed, and Bob blames "the effeminacy of modern times."

> Mankind are not satisfied with practising every art that can conduce to their own degeneracy; but they have likewise endeavoured to give a new and softer tone to animals. It was supposed that an intermixture between the genuine Terrier and the small Beagle would produce a very delicate variety, and unite the agreeable qualities of both. . . . The mixed breed has been reckoned much more elegant in form, more agreeable in manners, more beautiful in the variety of colours, than the pure of either.

• • •

THE QUESTION of why people choose to live with one dog rather than another finds no answer in science. What we suspect to be true may indeed be true: owners choose dogs that resemble them in temperament and appearance, or somehow "complete" them. Or it may be the case that owners want dogs that resemble babies. Or dogs that will in some way compensate for an attitude they lack or wish to project—aggressive dogs, huge dogs, sociable dogs. One might imagine that dogs with a bad image, such as those with behavioral issues or those prone to disease, would prove less popular in time, and would breed themselves out of the gene pool. Alternatively, one might suspect, and certainly hope, that dogs with the greatest longevity might prove to be the most popular: Wouldn't you want a dog that gave you the most reliable companionship, the greatest assistance and the promise of inseparable love to be in your life for the longest possible time? It appears not.

Until recently this evidence was largely anecdotal and unquantifiable. But in 2013, a quartet of social scientists from Brooklyn, North Carolina, Pennsylvania and Stockholm published a report suggesting dog owners consistently choose their companions based on strange, random and sometimes bad criteria.

The researchers began with a question they couldn't immediately answer. Why did the number of registrations of Irish setter puppies with the American Kennel Club, which stood at about 2,500 in 1961, jump to over 60,000 in

1974? And why did they fall back to about 3,000 by 1986? Could it be that a movie or popular television show was responsible? Was there a music or sports star who posed with Irish setters until they, and then their dogs, fell from favor? (The answer to these last two questions was "probably.") And how to account, from 1926 to 2005, for the changing popularity of other dog breeds registered with the AKC, from the consistent rise during this period of the Labrador retriever and the Yorkshire terrier, to the slow and sad decline of the Boston terrier, the American water spaniel and the greyhound? The changing role of the dog in society—from agricultural, hunting and sporting roles to that of domestic pet—may account for some of this, but there are also other factors at play. One factor that did not seem to be overly significant was the longevity, health or behavioral performance of the dog.*

The study investigated the relationships between breed characteristics and breed popularity by collating popularity data from the AKC, behavioral data from the Canine Behavioral Assessment and Research Questionnaire (C-BARQ), and longevity and health data from several

---

* The AKC registered more than 50 million dogs from 150 breeds in this 80-year period. It was assumed that the behavioral traits of these breeds (e.g., aggression or trainability) didn't change significantly in this time. The C-BARQ data examined the behavior and temperament of more than 9,000 dogs across 92 breeds, and were measured across 14 traits, including trainability, a fear of strangers, a dislike of unfamiliar noises and objects, a fear of separation, excessive attention seeking and aggression toward people and other dogs. As one might expect, the golden retriever scored high on only one of these: trainability. The Chihuahua scored high on most of them apart from trainability. And the Dobermann pinscher scored high when it came to chasing, excessive energy levels and attention seeking.

sources, not least the records of vets, dog hospitals and single-breed databases in the U.S. and U.K.

The researchers found no indication that behavior, health or longevity have been important in determining breed popularity. Rather the opposite, in fact: the most popular breeds had significant health problems (hereditary diseases such as hip dysplasia in Labradors, bladder stones in miniature schnauzers or von Willebrand disease too frequently afflicting the corgi, the German shepherd and the standard poodle), and possibly more behavioral problems. Conversely, there was no indication that breeds with more desirable temperaments, longer life expectancy or fewer inherited genetic disorders have been more popular than other breeds. Dogs that are considered harder to train, and suffer more from separation anxiety, have become increasingly popular. What does that say about owners, then, that health considerations are a secondary consideration when acquiring a domestic dog? And should dogs and owners be concerned that the researchers predominantly found, over time, "the kind of large and apparently whimsical fluctuations that are usually considered the hallmark of fashions and fads"?

Wally Conron is now in his late eighties. When reporters visit him in southeastern Australia he speaks mostly with remorse. He says he spotted the danger signs in the early 1990s, long before the designer mania we know today. Other, less experienced breeders produced labradoodles without the necessary care and knowledge to make them reliable assistance dogs; bred with limited awareness of hereditary health and behavioral problems, puppies would often appear slow, unpredictable and nervous. Conron produced thirty-one labradoodles himself, but by the time he retired (to what he calls his "little shoebox flat," confirming his limited means) there were thousands all over the world.

When asked whether he is proud of his achievements, he can only frown. "I've done so much harm to pure breeding and made so many charlatans quite rich," he told the Associated Press in 2014. "I wonder, in my retirement, whether we bred a designer dog or a disaster. Instead of breeding out the problems, they're breeding them in. For every perfect one, you're going to find a lot of crazy ones." He was horrified by the prospect of a cross between a poodle and a Rottweiler. "Why on earth would you do that?" he asked. It was a fair question, and it may have a simple answer: humans have ever sought both novelty and perfection. If we think we can somehow improve a dog, we will; if we can judge one dog to be somehow superior to another, we will turn the challenge into a profitable spectator event.

8.

How to Win a Ribbon

By the time I left Crufts, the annual British event billed as the world's greatest dog show, in the early evening of March 7, 2019, I was all kinds of nervous Rex. One enters Birmingham's National Exhibition Centre feeling fairly convinced that your dog is the best dog in the world, and one comes out feeling the size of an ant. Inside that vast and overrun pavilion is a master race, the finest and most sleekly groomed examples of the genus. Almost all pure breeds are represented, but only the best, the unforgiving aristocracy of those breeds. Not one dog has stinky breath or fewer than four names. Every nether region is as clean as a sink in a kitchen showroom. And many of these dogs have their noses so high in the air that they may as well be detecting life on Jupiter for all the attention they give dogs from the lower orders. Those top dogs are there to win, and if they can't win, they will humiliate, and stomp on your soul.

Apart from that, it was a lot of fun. During four days in March 2019, Crufts attracted about 160,000 humans and 21,000 dogs, and you could tell them apart because the humans were the ones who were drooling. There wasn't anything you couldn't buy to make a dog's life more like a human's, and to list even half the items would take up half a book, but it may be possible to begin with the products that are the most unnecessary.

We start with the Fit Fur Life dog treadmill (for over-fed dogs who have lost their zest; from £870 to £2,520, depending on size). Or how about Devine Spinning, the company that makes a plush knitted dog from your own dog's hair (may be used as cuddling aid or cushion). There is also the chance to buy a tweed dog bow tie from Hugo and Hudson, and a great many dog robes and smoking jackets. Is your dog wet? There are many new ways to dry him or her when a regular towel just won't cut it (the Doggy Bag, patent application 1310202.8, promises "Wet Muddy Dog goes in, Dry Clean Dog comes out," and the enrobed Cavalier King Charles Spaniel on the promotional leaflet looked like it wanted to stay in the Doggy Bag until it died). I was also interested in the Dogmatic head collar that replaces the traditional simple neck band with a more elaborate harness placed over a dog's neck and muzzle, something closely resembling a horse's bridle, designed to restrict a dog's tendency to pull on the leash and therefore "give you back control." Elsewhere there are many opportunities to hide or mask the realities of dog life, including the Bio-Enzyme Odour Management Formula from Aroma, promising to "eradicate completely the lasting malodours caused by urine, faeces or vomit on hard and soft surfaces all around the home and in the car." New puppy, anyone?

Dog shows have been selling dog products to dog people since the nineteenth century, much of them, as now, either largely superfluous, highly dubious or both. In 1889 a correspondent for the *Stock-keeper and Fanciers' Chronicle* attended the Windsor St. Bernard Show to find a company called Rackhams offering Distemper Balls, a "certain cure for everything." A firm named Astley was offering something that would be the last and only pill your dog ever needed, the ultimate in quackery, the "K-9 cure-all"; whatever your dog had, the K-9 would get rid of it. Bob Martin's offered "Aphrodisiacal Pills," presumably for the aging dog who had trouble finding next door's droopy setter as attractive as he used to, or vice versa. You could also buy a Victorian dog chain that was guaranteed to never tangle, or, should a tangle occur, to somehow untangle itself.

The main thing that's changed is the food on offer. For humans at Crufts it is primarily baguettes and hog roast, but for dogs the choice is double-pronged. For the traditionalist there is heavy meat: Platinum Natural Pet Food sells sachets with 70 percent fresh beef garnished with potato, vegetables, fruits, herbs and cold-pressed oils. Benyfit Natural offers a Pheasant Meat Feast: 80 percent wild British pheasant, 10 percent wild British pheasant bone, 5 percent ox kidney and 5 percent ox liver, designed to mimic a wild nutrient-dense diet from the time dogs lived in, or just outside, woodland. The company also makes goat, rabbit and venison meat feasts, and a seasonal pack of Turkey Christmas Dinner with turkey crown, parsnips, Brussels sprouts and cranberries. Food stalls nearby offer Meatiful, Ready Raw, Carnilove and Meatlove, all of them sounding particularly tempting with dimmed lights, and that's before Meatlove seals the deal with its signature "treat training sausage."

On the other side of the arena there are perfectly balanced meat-free options, many catering to gluten intolerance and celiac disease. Here you will find Yakers (snacks made from Himalayan yak milk), a Cold Pressed Fish Supper (with turmeric, linseed oil, yucca and apple cider vinegar), and Ocean Care hypoallergenic salmon and rice (enriched with copper). The transmission of nutrient care from human to dog is swift these days: we have an intolerance or allergy to something, or a faddish crush on something Californian, and our dogs have it too. Turmeric trace for dogs? Crazy to think we once thought bones would suffice.

When all your money's gone, you may permit yourself a brief interval to see actual dogs. The highlights of Crufts in the flesh bear only a passing resemblance to the highlights of Crufts on television. On television it is mostly perspiring fatsos trotting around beneath spotlights and anxious brows, but these big arena shows take place in a separate part of the NEC from the exhibition halls, and it is in these much smaller Astroturf rings where the trials, heats and junior competitions occur, the place where hearts are broken early. Only one dog can win each group (Gundog, Utility, Working, Toy, Pastoral, Terrier and Hound), and only one of these can win Best in Show, but winning anything at Crufts—even the third-place rosette in Ringcraft or Junior Handling competitions—is a major achievement.

Inevitably there are a large number of dogs who get packed away nightly in the Astra without anything, and on the drive back to East Sutton or North Weald through a freezing March night all those dogs and their owners must necessarily ask themselves what it is all for.

It is for the joy of dogs. It is for the joy of being with other people who are as crazy about dogs as you are. It's for the chance to find some piece of doggy merch that you never knew you needed, and for the knowledge that after Crufts there will soon be the Easter Classic at the National Show Centre in Cloghran, County Dublin, and the 195th Benched General Championship Show at the Royal Highland Showground in Edinburgh, and the 91st Annual Championship Show of the Bath Canine Society in Bannerdown, Bath.

Here are some other leading issues people and dogs who have never been to Crufts may want explained:

Q. Does the NEC smell really bad, and where do dogs go to the toilet?

A. It smells no better or worse than the average home with dogs. There are several fragrantly regulated sandy "run-off" areas on the sides of the halls where business may be attended to.

Q. Is the agility competition—the crazy activity where dogs have to run through PVC pipe tunnels, weave through a line of poles and charge at A-shaped ramps—as stupid and brilliant as it seems on television?

A. Yes, of course, but only when it goes wrong. It's like Formula One—you don't pay to see fast, you pay to see action. Sometimes, if one is very lucky, one sees the dog equivalent of a crash, a dog who says, *No way I'm jumping that, I'm just going to have fun!*

At Crufts this year that dog was Kratu. Kratu was a shaggy Romanian rescue dog, and he was so poor at agility that everyone who saw him wanted to take Kratu home and hug him. Peter Purves, the former *Blue Peter* presenter and regular Crufts commentator, introduced Kratu as "a large dog from Wood Green" which is technically not a breed yet recognized by the Kennel Club. In the words of Purves, Kratu was "trying to do a course," which is a bit like saying Harry Kane was "trying to do a goal," and it didn't actually turn out to be true. Kratu was trying to do a laugh, or trying to do a muck-around, but he had long ago decided to leave the course for another day.

He jumped the first hurdle well, but completely ignored the second one. He then had to negotiate the curly plastic tunnel. He went in okay, but he didn't come out okay; he just sat at the end of it licking his lips. He eventually popped out, and his handler Tess dragged him over the course by his collar, until he arrived at the next hurdle, which he jumped. But he refused the hurdle after that, running instead to the entrance of the ring. He then ran into the tunnel again, from where he once again refused to come out. He then turned around inside the tunnel, and reappeared at its entrance looking lovely.

Purves, who couldn't decide whether to continue reading from his notes or just laugh, continued reading from his notes: "This is a dog rescued in Romania from absolutely terrible conditions . . ." but then he gave up and said, assuming the voice of a dog, "It's nice in here, I like it in the tunnel!" After a few more completely random decisions by Kratu, Purves explained, "There is actually a proper route that they are supposed to take, and Kratu is not taking it, is all I can say." He also said that Kratu was the first Romanian rescue dog to be trained as an assistance dog, and he "can do all sorts, except he can't, er, follow instructions." Quite what these "sorts" were remained a mystery, although you probably wouldn't want Kratu to lead you across a busy intersection.

Q. How agile does a dog have to be to win the agility competition?

A. Extremely agile. The idea is to achieve the fastest speed over the course with no penalties. How may a dog incur a penalty? The Kennel Club brooks no crapulence, and imposes harsh demands. You will accrue five faults for failing to negotiate an object correctly. You will accrue five more faults if the handler deliberately touches either the equipment or the dog, or both (ten faults). Assisting a dog to complete an obstacle will cause immediate elimination. The refusal of an obstacle will cause five faults, and three refusals will cause elimination. A failure to attempt after a refusal will cause elimination. Taking the wrong course will cause elimination. Harsh handling will cause elimination. A dog wearing the incorrect style of collar will cause elimination. A dog judged to be out of control will cause elimination. Fouling the ring—i.e. urine elimination—will cause elimination.

Q. If, in a puerile mood, you had to select the most ludicrous names of recent Crufts champions, where would you start?

A. Bumblecorn Cats Nightmare
You Can Choose if Love or Fame
Plumhollow Top Hat

Roughshoot Fire and Ice with Baratom
Kiswahili Quantum of Solace
I Believe in Angels Oasis of Peace

Q. Does everyone love Crufts?

A. No. People have been complaining about the iniquities, anomalies and downright unfairness of Crufts since Crufts began, and probably before. The Victorians criticized both the corruptibility of the judges and the "monstrosities" they judged. Today, many feel uneasy with the extreme primping and objectification. Animal rights campaigners, most prominently People for the Ethical Treatment of Animals (PETA), don't like it because of the extreme breeding that it takes to be a champion, or even an entrant. "Contrary to the breeding industry's propaganda," PETA explains on its leaflets, the demands of Crufts and, by extension, all high-level dog shows, "continue to cause dogs to face painful and life-threatening genetic defects and diseases. Through an independent scientific report, the Royal Society for the Prevention of Cruelty to Animals (RSPCA) found that any changes made by the industry to breeding practices have been implemented too slowly and don't even come close to adequately addressing the serious welfare concerns. Physical attributes still dominate breed standards, and animals' health remains a secondary consideration." PETA also believes, with a logic that's hard to dispute, that Crufts exacerbates the UK's homeless-animal crisis: for every celebrated purebred dog that wins loving approval from those watching at home, figure on one new litter and one less visit to a rescue shelter.*

* In 2012 PETA made a mock poster mimicking the Victorian circus: "Annual parade of genetic freaks prone to disease and disability!" it proclaimed. "A Cavalier King Charles Spaniel born with a skull too small for its brain!" In 2018, the organization disrupted the Best in Show finale live on television to amplify its cause. Crufts organizers argued that the action itself endangered the safety of the animals, which PETA denied: "The dogs were in an arena with thousands of people, bright lights, music, and loudspeaker

The Kennel Club has a standard response to these accusations. It states that breeders should always adhere to the Animal Welfare Act (2006), and that prospective dog buyers should ensure they conduct their own thorough research and always buy from a responsible breeder. The independent report on pedigree dog breeding in the U.K. commissioned by the RSPCA and cited by PETA is by Dr. Nicola Rooney and Dr. David Sargan. Published in 2009, it provides a thoroughly useful survey, and concludes thus: "Welfare charities, veterinary associations, dog breeders and all other stakeholders must unite in using the latest advances in genetics and epidemiology to find a new model of dog-breeding practice. To say that there are numerous challenges facing the selection of healthy dogs in order to produce healthy offspring would not be an overstatement."

Q. Has there ever been a more acute and poignant display of inbred Englishness than at Crufts?

A. No. It may not be the biggest dog show in the world, (that would be in Leipzig, Germany, with almost four times as many dogs), or the longest established dog show (probably the Westminster Dog Show in New York City, founded in 1877) but in 2019 Crufts is certainly the only one with Mrs. Renée Sporre-Willes, the judge in the terrier group. Sporre-Willes attended her first Crufts in 1966, when what she calls "a small apricot poodle" won Best in Show. When asked what makes a Crufts dog extra-special today, she replies "a really typical head."

In 1937, the Irish poet (and borzoi owner) Louis MacNeice visited Crufts at Olympia for the magazine Night and Day. He was upset that the show, then in its seventy-sixth

---

announcements. They were made to prance around the ring—sometimes wearing choke collars. They were sprayed with styling products, poked, and prodded. The entrance of two young people holding signs is hardly likely to be the cause of their anxiety."

year, featured owners who were better behaved and less eccentric than he'd witnessed at a previous incarnation at Crystal Palace. He had no time for the people who complained that breeding for show purposes produces "an artificial type" of dog. Dogs lead such artificial lives, he argued, "that they may as well look artificial into the bargain." He admired the sheikh-like Afghan hounds, and the twenty-one Newfoundlands "walking about like sofas," although he had less time for dogs he judged to have lost some of their traditional characteristics (Great Danes had become too "gawky," and the Pekingese "vulgarly pretentious"). But what made Crufts Crufts? "Old ladies in all directions produced sandwiches from handbags, not to mention small but attractive bones. England is still herself, I thought . . ."

The one thing that doesn't make Crufts English is the fact that complete strangers are often to be seen talking to complete strangers: "You have a dog? I have a dog! What sort of dog is she? She's a really good dog and I bet you have a really good dog too . . ."

Q. How does a dog win Best in Show?

A. Technically the process is simple, and may be thought of as a pyramid. Local heats, then regional heats, then a competition at Crufts to win the best in breed, then a competition for that breed within its group (Gundog, Toy, etc.), then each group winner competing for Best in Show, the whole shebang taking place over several months before the concentrated period at the NEC over four days.

But to win at Crufts a dog is clearly not judged by the same criteria we may judge a dog at home, such as cuddleability and stench. A champion requires an alchemical mix of fine breeding, conformity, a calm but vivid temperament, relentless primping and a secret sauce even the judges can't easily define. A dog needs to match its breed standard to an inch of its life, but it also needs sparkle, ring presence and a perfect rapport with its handler. It is not good enough, in other words, just to be a good dog.

Not so long ago it *was* possible to be merely a good dog and win Crufts. Up until 1967, when "Cruft's" was held at London's Olympia exhibition center and still had an apostrophe, any dog could enter on opening day with wild fantasies of being there on closing night. All sorts would turn up with all sorts of things wrong with them that their owners had deemed normal, including inexplicable body lumps and distemper. These days, the champion dogs who do not win Best in Show have jumped so many preliminary hurdles to get there that they and their breeders may still be regarded as champions.

That said, several of those I spoke to had a resigned air about them. Teresa Barker, for example, who has been breeding and showing Chihuahuas for more than thirty years, told me that although her dogs, both smooth-coated and long, have occasionally triumphed in their class, beating out twenty other Chihuahuas, it was "impossible" to win the big prize. At Crufts there could be more than two hundred Chihuahuas on show, "but the biggest breeders keep their very best dogs and they will not sell you their puppies. I've been on waiting lists for a very long time." Recently, Barker thought she had bought a potential smooth-coated winner from a breeder in Central Europe, but when the dog arrived, she discovered a fault that wasn't visible on the seller's video. "They hid it from me. They're meant to have very straight front legs, so that their movement is in a straight line, but hers have got a slight curve in them which you can hide when they're jumping around or standing on a table." Barker, who lives in north London, shows her dogs all over the country, usually four at a time, but she told me that recently she and her friends have begun to question why they do it, so tough is the competition and so considerable the expense. Also, her second husband isn't such a great fan of Chihuahuas, which, given that Mrs. Barker has ten, does occasionally push his love for her to the limit. (Her first husband loved dogs, however, and came with his own German shepherds.)

To get an idea of what judges are looking for these days, one may consult the official reports from regional shows printed in *Our Dogs*. These are shows you need to win to have a chance of appearing at Crufts. So at the Maidenhead & District Canine Society show, for example, which in February 2019 was won by a four-year-old Slovenian champion papillon (a small, highly coiffed spaniel-family lapdog) named Whip Honey Double Smash, an onlooker unfamiliar with the breed may have remarked how beautiful he was, and how his striking black eyes and ears made him look like a masked superhero. But to Toni Jackson, who judged him Best in Show, he was "a delightful showman who is very alert and attentive to handler, putting in a fabulous performance. Presented in super condition, both coat and muscle. Lovely expression from correct skull and wide set feathered ears, with dark medium eye. Well constructed, moved true in all directions, with lively gait."

Clearly this was no "large dog from Wood Green." Reserve Best in Show at Maidenhead went to a two-year-old Siberian husky named Jacalous Catch Me if You Can for Vukasin. Judge Jackson's eye was enamored with the fact that Jacalous Catch Me if You Can for Vukasin had "excellent balance and lack of exaggeration." In addition, Jacalous Catch Me if You Can for Vukasin displayed a lovely temperament and "a typical fox-like head, with lovely almond eye, giving bright expression." She was "well boned and coated without being overdone" and had "well laid back shoulders and good bend of stifle [knee] giving complementary angulation." So, well done (but not overdone), Jacalous Catch Me if You Can for Vukasin.

There is unmistakable delight in these descriptions, and from other show reports. "Enjoyed judging this delightful breed," Toni Jackson said. Another judge, Bridget Harris, adjudicating at the London Cocker Spaniel Society, also found it an honor and a privilege to be asked. She "very much enjoyed going over some quality dogs." although she did notice that "a few exhibits could have had better dental care." These judges, and the people whose dogs they judge, were once collectively known as "the dog fancy," although the phrase is now somewhat antiquated. Quite as popular these days is the description "doggy." "You're not really *doggy*, are you?" a doggy person may ask another person who isn't carrying chicken treats or covered in hair.

Q. Can you buy a dog at Crufts?

A. Not exactly. But after you visit you will have a much clearer idea of the sort of dog for you. The most useful part of

Crufts is not the agility or obedience displays, but the booths around the perimeter of the halls containing one or two of almost every breed it is possible to own (about two hundred in all). If some terrible fate was to befall the universe beyond the NEC, it would be possible to preserve and regenerate almost the entire purebred dog world just from these fine specimens. Fun too.

The dogs in the booths are displayed in alphabetical order according to breed, which may lead to some sudden changes of devotion. You may arrive with your heart set on a longhaired dachshund and end up falling for a Cirneco dell'Etna, which will need a bit more exercise, a larger couch and a different language. Or you'll go in looking for a giant Newfoundland (so big it needs two booths: W25/26) and go to pieces over next door's medium-size Portuguese water dog (W27). Within their booths, the dogs are described according to exercise needs (from less than half an hour a day for the Japanese chin to more than two hours a day for the giant schnauzer), their expected longevity (from less than ten years for the mastiff to more than twelve for the giant schnauzer) and whether they are best suited to town or country. The Newfoundland is decidedly rural, while the Russian toy prefers the urban mews. The giant schnauzer is happy in both environments, so is extremely suitable for city dwellers with a weekend retreat.

In a bid to show it is not entirely humorless or out of touch, in 2012 the Kennel Club launched Scruffts, a competition running in parallel with Crufts but specifically for mutts. This rewarded, in unwoke fashion, the "most handsome" crossbreed dog, the "prettiest" crossbreed bitch, a "golden-oldie" crossbreed (ages eight and above), the best crossbreed rescue dog and the crossbreed deemed "the child's best friend." It marks a welcome change from an earlier era. In 1932, an issue of *Country Life* carried an advertisement imploring its readers, "Don't ever buy a mongrel—Get a *good* dog. Breeders sell them very reasonably. Make inquiries about any breed you fancy from exhibitors at Crufts Dog Show next week."

CHARLES ALFRED Cruft did not invent the dog show, but he was a showman, so he swiftly made it his own. In 1891 he created the first dog show that even people without a dog had heard of. He brought with him the hoopla of the circus and the variety of the music hall, and by the time his first event ended at the agricultural hall in Islington, he had made the dog show industrial.

He was the son of a London jeweler, and his earliest (and only) stated ambition was to make a success of his life. His first professional connection with dogs came when he worked in his twenties for James Spratt, one of the first companies to engineer food specifically for animals. Spratt's biscuits consisted of grains, beetroot and a vague sort of beef fluid (probably blood), initially marketed as Spratt's Patent Meat Fibrine Dog Cakes. Cakes they were not, or at least not in the modern sense of sponge and moistness. The items were based on the indestructible dry snacks endured by sailors at sea, and they swiftly became a popular choice for the landed gentry, who fed them to their hunting dogs. Savvy marketing linked the product with successful entrants at dog shows, the way Purina and Iams did years later; one fed the other.

Cruft's skill was in sales and marketing. His promotional work for several dog societies led to his management of the dog sections at the Paris World's Fair in 1878. In London he ran shows for spaniels and terriers, with almost six hundred entrants attending the Old Royal Aquarium in Westminster in 1886. The show was extended to toy dogs in 1890, and to all breeds the year after (the first show offered 2,437 dogs from thirty-six breeds).* Just before the outbreak of war in 1914, more than four thousand dogs entered, and Cruft's had a worldwide reputation, even though the dogs and the judges were regarded as substandard and corrupt by older institutions like the Kennel Club.† Cruft dismissed all criticism with the bravado that had become his trademark, although he wasn't beyond being duped himself. The early manipulation of the dog through crossbreeding was at its height, and in his quest for the perfect dog in all categories, Cruft encouraged this new rough science. His shows are credited with the rapid

---

* These days the phrase "toy dog" summons images of B-listers with handbag or teacup dogs; originally the term merely meant "small," and generally referred to lapdogs (who may, or may not, have been just as pampered by Victorian ladies as the current crop).

† The Kennel Club was established in 1873, not least to regulate the showing of dogs and control the Victorians' mania for new breeds. One entry in its "Code of Rules" produced the following year banned the sort of poor dog who may have been permitted previously: "No dog shall be qualified to compete . . . who is suffering from mange or any other form of contagious disease." Charles Cruft died in 1938, by which time around ten thousand dogs attended his show annually. The Kennel Club bought the rights to the show from his widow, Emma, in 1942.

popularization in Britain of imported novelties such as the borzoi, the saluki, the basenji, the Rhodesian ridgeback, the Pekingese and the Lofoten puffin hound (also known as the Norwegian lundehund). The Siberian eel hound might also have had its day in the sun in 1931, had Cruft not been alerted just before the show opened that its entry was a prank, named after a fictional dog in *Very Good, Jeeves!* by P. G. Wodehouse.

Cruft had many forerunners, not least Charlie Aistrop and Jemmy Shaw, who ran competitions in the mid-1850s, often in the backyards of pubs, primarily for dogs who could prove themselves as ratters. (The winners were then sold to the highest bidder with the most rats. Shaw also did a neat trade in other rat chasers, including cats, ferrets and mongooses.) The earliest official best-in-breed show is usually credited as taking place at the town hall in Newcastle upon Tyne in 1859, with judges awarding prizes (of shotguns) for the best "version" of a dog irrespective of its usefulness or ability, but only pointers and setters were represented, and there were only sixty dogs entered.*

---

* The show in Newcastle was, according to a subsequent report in the Kennel Club's *Stud Book*, "held . . . at the suggestion of Mr R Brailsford." Richard Brailsford, a gamekeeper and dog trainer (or "breaker"), had previously written in the *Field* that too many dogs were "weeds, wastrels and mongrels," and too many were being crossbred for commercial gain. What was needed was a way of improving and maintaining purebred dogs through competition. So who, one wonders, won the first prize in the pointer category of that first official show? Step forward Bang, owned by Mr. R. Brailsford. And who should win in the pointer bitch division at the next big show held in Birmingham a few months later and organized by Mr. R. Brailsford? Only Mr. J. Brailsford. And what about Leeds in 1861? A retriever prize goes to Mr. H. R. Brailsford and his dog Windham, while in

Three years later a show in Islington attracted the attention of a correspondent for *All the Year Round*, the weekly journal edited by Charles Dickens.* "Some prizes are to be won by size, by depth of chest, by clean finish of limb, and symmetry of points," he noted, referring primarily to the classes of setter, retriever, greyhound and pointer.

> Meanwhile, to be bandy, blear-eyed, pink-nosed, blotchy, under-hung, and utterly disreputable, is the bull-dog's proudest boast. The bloodhound's skin should hang in ghastly folds about his throat and jaws, with a dewlap like a bull. The King Charles's spaniel wears a fringe upon his legs like a sailor's trousers, and has a nose turned up so abruptly that you could hang your hat upon it if it were not so desperately short.

The writer went on to expose a painful anomaly.

> A dog is a great bore; he howls in the night; he is tiresome to feed; he wants to go out when you have somewhere to go on business and cannot take him, on which

---

the pointer class it's all about Mr. W. Brailsford's Moll. Is there, perhaps, a Mr. F. Brailsford in the Brailsford clan? There is, and he's among the prize-winners for retrievers at the Chelsea show, where his Shot is joined by Mr. J. Brailsford's Bell. Astonishingly, Mr. J. Brailsford's Jip doesn't actually win any of the top three prizes at the First Great International Dog Show held at the agricultural hall in Islington in 1863, but a riot is averted when he is awarded an "Extra Prize" in the retriever class.

* The correspondent was once assumed to be Dickens himself, but scholars now usually credit John Hollingshead, who later went on to have great success as a theater manager.

> occasions to see him with his head on one side looking after you as you shut him up is enough to break your heart; he is disliked by your friends whose carpets he impairs and whose cat he frightens; he is liable to be stolen, and to catch distempers and other diseases—in short, he is altogether a heavy handful, but still, if any one were to offer me one of those real old-fashioned spaniels, I hardly think I could refuse the gift.

The article was entitled "Two Dog Shows," and the second one was in permanent residence a few hundred yards north. This was a less glamorous exhibition, enclosed by three walls and a screen of iron cage-work. "As soon as you come within sight of this cage some twenty or thirty dogs of every conceivable and inconceivable breed rush towards the bars, and, flattening their poor snouts against the wires, ask in their own peculiar and most forcible language whether you are their master come at last to claim them?"

It was an innovative London dog pound, founded eighteen months earlier and already responsible for taking in more than a thousand lost souls, "the poor vagrant homeless curs that one sees looking out for a dinner in the gutter, or curled up in a doorway taking refuge from their troubles in sleep." The writer concludes, with all the humanity one would expect from a Dickens publication, that

> this is the kind of institution which a very sensitive person who had suffered acutely from witnessing the misery of a starving animal would wish for, without imagining for a moment that it could ever seriously

> exist. It does seriously exist, though. An institution in this practical country founded on a sentiment . . . evidence of that hidden fund of feeling which survives in some hearts even the rough ordeal of London life in the nineteenth century.

The sentimental and noble woman responsible for this institution was Mary Tealby, and she named it the Temporary Home for Lost and Starving Dogs. It relocated to southwest London in 1871, when it became the Battersea Dogs Home.

In many ways, the nineteenth century witnessed the assembly of an entirely new dog, both in the flesh and the imagination. It was the century of regulation and standardization, of fear and celebration. The Kennel Club was established in 1873 to oversee the rules and financial management of dog shows, and to regulate the consistency of pedigrees. Much of the dog news from nineteenth-century newspapers seemed at odds with itself: for every spike in rabies there was a new dog show; for every outcry about the number of strays on London's streets there was an improved ladies' lapdog for the front parlor. Dogs became aspirational objects, the traditional Victorian pyramid at its most steadfast: you welcomed a bloodhound into your middle-class home, and you became just a little bit more landed gentry.

The scholars Michael Worboys, Julie-Marie Strange and Neil Pemberton have credited the Victorians with "inventing" the modern dog, with one dog in particular, a large pointer called Major, standing four-square-legs at the center of it all, an unwitting canine patient zero. In 1865, Major was described by a surgeon-turned-sportswriter in the *Field* to within an inch of his life. No dog had received such close scrutiny before: he was drawn and quartered, and then quartered thrice more. Divided into sixteen parts, each had a score: within the category "frame and general symmetry" there were points given for loins (up to a score of seven), hindquarters (six), shoulders (five), chest (four) and symmetry (three). The pointer was also given aspirational guidance in verbal form. "The lips should be well developed . . . nose long, broad and square in its front outline: that is to say, even jawed, not pig-snouted."

This close typing and categorization—elsewhere so visible among Victorians in their pursuit of fossils and stamps—was swiftly extended to other breeds as the Kennel Club developed its *Stud Book*. From this pointer onward, no purebred dog, and certainly no prizewinner at Crufts, would ever be allowed just to be a dog in the same way again. The *Stud Book* swiftly became the bible, the final word in how a dog should look and how it performed in competitions. And so it remains: the 2019 volume runs to more than one thousand pages. But other books were and are available, and it is through these that we may define how we have truly valued our dogs through the centuries.

9.

Dog Story

The literature of dogs—that is to say writing about dogs, by dogs, and for dog owners—goes back, as you might expect, to the dawn of writing itself. Longer even: as we have seen, dogs were a weighty part of tribal myth, a focus around the campfire. We were not walking or hunting alone, and there would be no reason to exclude our friends from our stories: indeed, as the following pages make clear, often our friends *were* our stories. The first book to focus entirely on dogs appeared nearly five hundred years ago, and a trickle soon turned into a flood. Dogs swiftly became a staple of childhood literature; the word "dog" is often one of the first we recognize and pronounce as children. The trickiest issue in any brief consideration of the subject is not where to start but where to end.

For there is a logical place to start. So many authors have made the dog the subject of some of their finest writing, and in so doing have shown us a new animal, or at least a familiar animal in a new light; such writing cannot but help illuminate all the intricacies of our bond. We should take into account Jack London, Charles Dickens, P. G. Wodehouse, J. R. Ackerley, P. D. Eastman and Plutarch, but it should inevitably begin with a Woolf.

Was there anything Virginia Woolf didn't know about dogs? She surrounded herself with dogs, she wrote an entire novel about a dog, dogs were involved in the wooing of her secret lover and one of her very first published articles was an obituary of a dog.

But her relationship with dogs did not begin well. At the age of nine Woolf wrote in her family's self-produced newspaper *Hyde Park Gate News* of being attacked by an animal known as the Big Dog, who bit her cloak and pushed her against a wall. In contrast, her youthful obituary of her family dog Shag was affectionate and moving, though it didn't skimp on the downside.* She describes him as a bit of a biter who, with so much rabies around, was encouraged to wear a muzzle.† Shag was also a bit of a snob, as "he seldom went for a walk without punishing the impertinence of middle-class dogs who neglected the homage due to his rank." Woolf reports that Shag became increasingly

* The Shag obituary, which was among her earliest published works, appeared in the *Guardian* in 1904 (not the newspaper, but a friend's journal for the clergy) when Woolf was twenty-two and eager for any payment to facilitate her financial independence.

† By 1895, 672 cases of rabies had been reported in England, and the debates over muzzling were fierce (vets and the pet journals advocated it; pet owners complained it invaded their privacy and personal liberty; the government made it compulsory).

autocratic, so that the thought of him having a master and mistress became untenable, and everyone instead became his aunt and uncle. But among his own class Shag is positively clubbable: "I can see him smoking a cigar at the bow window of his club," Woolf writes, "his legs extended comfortably, whilst he discusses the latest news on the Stock Exchange with a companion."

What else do we know about Shag? He was gray and shaggy. He liked a good shag. He had a collie's head and Skye terrier legs, but he was decidedly not the pure terrier they thought they had bought to rid them of rats in their summer house in St. Ives. Alas, he also possessed a temper that didn't work out well for him. He once bit a visitor who had the temerity to call Shag what Woolf recognized as a "contemptible lapdog title": they called him Fido.

Shag had the sort of class issues and sense of entitlement one sees a lot in the older breeds (old money versus new), and his inability to make allowances for a changing world led to his banishment. Woolf describes the arrival at her family home one day of a puppy named Gurth, an Old English sheepdog in traditional white and gray. Old Shag was not pleased, and he behaved pathetically. Seeing that the new dog loved to offer his paw, Shag tried the same trick, but it was forced and stiff, and few thought the aging dog was cute. Shag then tried another trick: he went for the new dog's throat. The two fought like Itchy and Scratchy. "When at last we got them apart," Woolf writes, "blood was running, hair was flying, and both dogs bore scars."

So Shag was sent to a servant's home in Parsons Green to spend his final years. Poor old Shag. Gurth the sheepdog went on to hold a special place in V.W.'s life, becoming an unwitting but invaluable part of the Bloomsbury Group, accompanying Woolf on outings to the London Library and even to concerts, where he once howled his own bass accompaniment to a song and had to be removed "with haste." For Woolf, Gurth's faithful presence resonated widely: she began using "sheepdog" as a term of affection for her sister Vanessa, and she extended the usage to close friends.*

But then one winter's night, a sort of lovely miracle. A barking was heard outside the Woolfs' family home, and when the door was opened, "in walked Shag, now almost quite blind and stone deaf . . . and, looking neither to right nor left, went to his old corner by the fireside, where he curled up and fell asleep without a sound." He was near the end. He met his fate shortly afterward at the ragged wheel of a hansom cab, a dramatic demise regarded as a

---

* The fact that Shag was replaced by an English sheepdog even shaggier than he was is clearly the key point here, and suggests to me that this may be the ideal time to relate one of the oldest original shaggy dog stories, the story that may well have inspired the phrase "shaggy dog story" in the first place. One day, in late Victorian England, a person sees a notice in *The Times* explaining that a wealthy landowner is distraught at the loss of his Old English sheepdog, who had been missing for ten days. The story in the paper describes the dog's coloring, his name and his all-round shagginess, and the gentleman reading the report is so moved by the loss of this dog that he decides to look for him. He fails: the poor dog is gone. But then he has another idea. He will find a sheepdog the same age and size as the lost one, and give that to the owner instead. It's a long search. A really, really long search. Eventually he is successful, and turns up with a beautiful dog at the landowner's country pile, only to be greeted by the butler. The butler looks down at the dog by his feet and says, "The last one was shaggy, but he wasn't *that* shaggy."

blessing. But he had returned home, loyal to the end, to say goodbye.

Perhaps no one should have been surprised when, thirty years after Shag, the now established writer produced *Flush*, her potent novella of another deceased dog, this one the fabled property of the Victorian poet Elizabeth Barrett. Before Hollywood made movie stars of Rin Tin Tin and Lassie, Flush was one of the most famous dogs in the world, a savior of souls and an inspiration to the lonely. Barrett wrote florid verses about him ("Like a lady's ringlets brown / Flow thy silken ears adown"), but it was Woolf's novel that secured his fame. The book offered a gimmick many must have thought would never work, but somehow struck a chord: a faux biography of a dog, sometimes truthfully reconstructed from the evidence, sometimes fantastically and empathetically enhanced with what Woolf imagined to be the dog's perspective.

It wasn't written *by* the dog, as has sometimes been assumed by those yet to read it (imagine the reception to *that*: dear old Virginia surely carted away in a van). Rather, the dog's point of view has been taken into account, although this alone was too much for some: critics damned the project as frivolous, and at times even Woolf appeared to disown *Flush* as unworthy of her reputation. In fact, it is a rich and rewarding book, especially for dog lovers. It's an exemplary account of a dog's ability to transform a single life. And not for the last time, a dog saw the world as a human would.

Woolf uses Flush to write about the Barrett–Browning courtship, and to describe how a new relationship may cause jealousy in an older one (surely shades of Shag).

Flush is the secret observer of the couple's elopement to Italy, and, when the dog gets kidnapped from Wimpole Street to live in a dive in Whitechapel, a smart way for the author to consider class divisions and social inequality. In her introduction to the Penguin Classics edition of the book, the scholar Alison Light called the work "a Woolf in dog's clothing," which is both funny and true. Flush himself, oblivious to the debate about his worth in his chronicler's oeuvre, is just a quarrelsome skit of happiness.

Flush was written not long after Woolf had finished *The Waves*, and it is as delightful as its predecessor was onerous. The book carries many examples of the love that refracts from dogs to humans and back again, and it doubles up as therapy for the author. Woolf had already experienced much personal tragedy and mental despair by the time *Flush* was published by her and husband Leonard's Hogarth Press in 1933, and she regarded it as she regarded dogs as a whole—with particular sentimentality. For her, "a dog somehow represents—no, I can't think of the word—the private side of life—the play side."*

But Flush has also been read as feminist allegory, a strong, searching figure in a teeming city, particularly

* Flush was the cocker spaniel delivered to Elizabeth Barrett as a six-month-old puppy in 1841. The poet was bedridden in Torquay at the age of thirty-five—a dark, closed-curtain malaise both physical and mental, and typically Victorian—but the arrival of a dog, a gift from her author friend Mary Russell Mitford, brought an uncharacteristic period of relief and abandon. In Barrett's letters, Flush "dances and dances & throws back his ears almost to his tail in Bacchic rapture. More than once he has lost his balance & fallen over." Elsewhere, the little dog "shines as if he carried sunlight about on his back." Oh, how I wish someone would one day write like that about me.

happy when uncooped, roaming Oxford Street and elsewhere, alive to the world. "The whole battery of a whole London street on a hot summer's day assaulted his nostrils," Woolf writes. "He smelt the swooning smells that corrode iron railings; the fuming, heady smells that rise from basements—smells more complex, corrupt, violently contrasted and compounded than any he had smelt in the fields near Reading; smells that lay far beyond the human nose." And this continued, with Flush in his own Arcadia, the smells of the world ruling his life the way they do with all dogs, "until a jerk at his collar dragged him on." Which inevitably leads to a question for all harried owners everywhere: Are you that jerk?

Away from her public writing, Woolf used her own dogs as symbols of erotica. When her lover Vita Sackville-West left for a trip to Persia in 1926, Virginia implored her to remember both her and her mixed-breed Grizzle (a successor to Gurth). "These shabby mongrels are always the most loving, warmhearted creatures," Woolf wrote. "Grizzle and Virginia will rush down to meet you—they will lick you all over." On her way back to London, Vita remarked, "This will be my last letter. The next thing you know of me, will be that I walk in and fondle Grizzle." (Nurse, the screens!)

And then there was the beautiful Pinka, the erotic purebred spaniel that Sackville-West had given Woolf as a present at the height of their affair. Virginia kept Vita updated on their growing devotion, not least on Pinka's not inconsiderable achievement of literally eating her husband's homework. 'Your puppy has destroyed, by eating holes, my skirt, ate L's proofs, and done such damage as could be done to the carpet. But she is an angel of light. Leonard says seriously she makes him believe in God—and this after she has wetted his floor 8 times in one day."

Dogs who eat proofs may merely be extending their role as protectors of literary taste. Writing in the same decade, John Steinbeck faced a similar dilemma. In May 1936 he wrote to his agent Elizabeth Otis that his Mexican setter puppy (whom he called Toby, like many of his dogs) had been left alone one evening and had "made confetti of about half my manuscript book." The book was *Of Mice and Men.* "It sets me back," he reported. "There was no other draft. I was pretty mad but the poor little fellow may have been acting critically."*

THE MOST cogent literary thoughts about dogs began where you'd expect—in temples and vanished libraries, with the Greeks.

Pythagoras won every dog's gratitude when, according to Xenophanes, "he passed by as a dog was being beaten, and pitying it, spoke these words: 'Stop, and do not beat it; the animal has the soul of a friend; I know this, for I heard it speak.'"

His contemporary Xanthippus (the father of Pericles) owned a dog with the soul of a friend. Plutarch relates that

---

* Steinbeck clearly knew his dogs. Of his French poodle Charley, he once wrote that he was born on the outskirts of Paris, and although he knew a little poodle-English, he only responded promptly to commands when issued in French. "Otherwise he has to translate, and that slows him down."

when, in 480 B.C., the Greeks set out to fight the Persians in the Battle of Salamis, their weeping relatives and their weeping animals were to be seen at the Athenian harborside saying farewell. Their chances of returning from the naval conflict were slim, and everyone knew it, but for some the prospect of disaster was just too much. And so it was, relates Plutarch, that the dog of Xanthippus "could not endure to be abandoned by his master, and so sprang into the sea, swam across the strait by the side of his master's trireme [an ancient galley with three rows of oars], and staggered out on Salamis, only to faint and die straightaway."* The poor creature supposedly lies on the island beneath the pile known locally, and rather unappealingly, as "Dog's Mound." Xanthippus survived the battle and lived for a further five years, and how lonely those years must have been without his friend.

It is inevitable that almost all writing about dogs—the most affectionate or angry sketch, the fleeting diary entry, the most convoluted novel—says as much about humans as it does about their animals. Dogs direct our empathies: a reader cannot love a character who kicks a dog, but a reader will swiftly welcome someone who rescues one, or halves their own rations so their dog won't starve. And a plot will often turn on a dog's loyalty.

So we can't leave the Greeks without at least a fleeting mention of Argos, who almost gives away the true identity

* One makes the assumption that when Xanthippus's dog swam out to the island of Salamis, about ten miles west of Athens in the Saronic Gulf, he did so purely out of blind loyalty toward his master, and not because he thought he was swimming out to a supremely large choice of salamis.

of Odysseus as he returns to Ithaca in disguise after his ten-year journey back from Troy. Homer invests Argos with a catalog of ailments—he is weary, he is flea-ridden, he lies amid cow dung—but still he recognizes his old loving master after twenty years. (Of course he would—probably by smell.) Only then, with his last act of love, does Homer permit Argos to pass into "the darkness of death."*

THE FIRST published work devoted entirely to dogs appeared in 1576. *Of Englishe Dogges, the Diversities, the Names, the Natures and the Properties* was written in Latin by John Caius. Caius was the physician to Edward VI, Queen Mary and Queen Elizabeth, and so had seen many of the earliest domesticated aristocratic dogges at close quarters, not to mention many hunting dogges in the field. His treatise reads like an anthropologist's report of exotic creatures on remote islands.

In the popular translation by Abraham Fleming, Caius identifies three distinct types of dogges. A currish kind (bad-tempered and angry), a gentle kind (used for hunting) and a homely kind ("apt for sundry necessary uses"). The gentle/hunting kind was useful for securing English

---

* And perhaps we shouldn't completely ignore Aristotle and his *Historia Animalium* from the fourth century B.C., speculative as it is. "The males normally lift the leg for passing urine at the age of six months, though some do it later, at eight months, and others before six months; as a general statement we may say they do it when entering on the period of exercising their full powers." And then there's the less equivocal observation in the same work that "a dog's age is told by inspecting its teeth: young dogs have sharp, white teeth, older dogs black blunt ones."

hart, weasel, lobster and polecat, and was divided further into subcategories, each with a particular attribute, "the first in perfect smelling, the second in quick spying, the third in swiftness and quickness, the fourth in nimbleness, the fifth in subtilty and deceitfulness." Caius then focuses on certain breeds. Setters, sheepdogs, terriers and greyhounds all feature, as do spaniels, with a measure of astonishment at the charm they exert over their owners. "It is a kinde of dogge accepted among gentles, Nobles, Lordes, Ladies, who make much of them . . . that they will not only lull them in their lappes, but kysse them with their lippes, and make them theyr prettie playfellows."

The index promises mucheth. "Butchers dogge why so called" scrolls back to the description of a dog that is forever chasing and protecting his sheep.* "Mooner why so termed" turns out to be a dog who's up all night annoying everyone by baying at the moon. Other entries are more self-explanatory: "England is not without Scottish Dogges"; "Gasehound sometimes loseth his ways"; and "Fisher dogge, doubtful if there be any such." There are also examples of faithfulness—dogges with tailes that doeth wagge at the verye sight of the master—but also of faithfulness turning sour on occasion of errant dogge who biteth.†

---

* The explanation of the phrase "butcher's dog" has since changed. In a dictionary of slang from 1859, it was redefined as "to lie by the beef without touching it; a simile often applicable to married men." The phrase "fit as a butcher's dog" can mean either a very thin dog, because he is forever dodging the greedy butcher's machete, or a very fat dog, because he's just lying around all day eating sausages.

† The literature of the Middle Ages concentrates on aristocratic hounds or gun dogs. (One notable exception: Chaucer's Wife of Bath says of a woman's

In the half-millennium since John Caius, the flood has not abated. So numerous have been the books about dogs—the breeding guides, the purchasing guides, the training manuals, the scientific reports, the fictional heroes, the scatological memoirs, the sentimental journeys—that inevitably one hesitates before joining their number. But such is the joy of dogs—their variation, idiosyncrasies, tragedies, solaces and bottomless well of unwavering love—that to restrict the flow would seem somehow unnatural, on a par with suddenly ceasing to write about love.

Everyone finds their favorites. My own is odd enough: John Galsworthy, and nothing to do with *The Forsyte Saga*, a firm favorite of my grandmother's. It is his sheer love of dogs that wins through, much of it in his ephemera. Writing in his essay "Memories" (1912) he asks, "If a man does not soon pass beyond the thought 'By what shall this dog profit me?' into the large state of simple gladness to be with dog, he shall never know the very essence of that

---

over-affection for a man, "as a spaynel she will on him lepe"). Shakespeare frequently made the dog his metaphorical pet, albeit an aggressive one. Mark Anthony entreats Caesar's spirit to cry, "Havoc!" and let slip the dogs of war. Shylock observes of his enemies, "Thou callest me a dog before thou hast cause. But since I am a dog, beware my fangs." Shakespeare's plays contain twenty-seven references to "cur" (an unpredictable, untamed dog), and almost as many to unpleasant barking. Dog insults are everywhere: the phrase "whoreson dog" appears both in *Cymbeline* and *King Lear*, Cleopatra addresses Seleucus as "slave, soulless villain, dog!" while Richmond bays over Richard III's limp body with "the bloody dog is dead." The nearest Shakespeare gets to a compliment is offhand: "I had rather hear my dog bark at a crow," Beatrice teases Benedick, "than a man swear he loves me." "William Shakespeare Hated Dogs" a headline stated not unreasonably in *Psychology Today*.

companionship which depends not on the points of dog, but on some strange and subtle mingling of mute spirits."

Elsewhere, Galsworthy recounts a trip to Waterloo Station one day with his wife, Ada, to collect a black spaniel sent by friends in Salisbury. They take him home. "When he just sits, loving, and knows that he is being loved, those are the moments that I think are precious to a dog; when, with his adoring soul coming through his eyes, he feels that you are really thinking of him."

He returns to his theme in a letter to *The Times* the following year, comparing the human sentiment toward dogs to the emotion one feels toward children. In other words, we afford them a place in our lives that they deserve. They are "by far the nearest thing to man on the face of the earth . . . the one dumb creature into whose eyes we can look and tell pretty well for certain what emotion, even what thought is at work within; the one dumb creature which—not as a rare exception, but almost always—steadily feels the sentiments of love and trust."

Less of the "dumb," perhaps. Could a dumb creature turn author and write a book such as *Thy Servant a Dog* (1930), a tale of mild misbehavior among dogs, a rat and a cat told by a black Aberdeen terrier named Boots (and Rudyard Kipling); or *The Life and Opinions of Maf the Dog, and of His Friend Marilyn Monroe*, a four-paw celebrity account by the dainty Maltese originally owned by Virginia Woolf's sister Vanessa Bell (written with Andrew O'Hagan); or *The Last Family in England*, in which an erudite Labrador learns to keep a troubled domestic unit together (with Matt

Haig); or *Timbuktu,* written by the canine wonder that is Mr. Bones (assisted by Paul Auster, who was inspired by a dog called Ollie he saw using a typewriter); or the viciously comic parable *The Heart of a Dog,* in which a stray becomes semi-human and learns to write like Mikhail Bulgakov; or really any volume in the Chet the Dog mystery series ghostwritten by Spencer Quinn, not least *A Fistful of Collars, Scents and Sensibility, The Sound and the Furry, To Fetch a Thief, Heart of Barkness* and the rewardingly suggestive *A Cat Was Involved*?*

And then there's my childhood favorite: *Go, Dog. Go!* by P. D. Eastman, who was a friend of Dr. Seuss. The joy to be derived from *Go, Dog. Go!*, beyond the idiosyncratic punctuation of the title, is twofold: the iris-searing illustrations and the rhythmic architecture of the ludicrous text. The style is snappy, being a series of short conversations between a fancy-pants poodle and a lovable if irritating bloodhound-style mutt. The story could never claim to run deep, but it runs surreal; to a child, the universe of *Go, Dog.*

---

* A cat *was* involved! Quinn and Chet's blog isn't bad either, containing as it does such jokes as "What do you call a zoo with no dogs? A shih-tzu." (Not to poo on your snack, but surely a zoo with no dogs is the best sort of zoo you could have?) My childhood library is packed with dogs who bark or howl or whimper, but none have the force of *Hairy Maclary From Donaldson's Dairy* by Kiwi Dame Lynley Dodd. It's nonsense on the surface, and nonsense below the surface too, but it does provide a useful beginner's guide to the confusing thicket of multiple breeds. There is the old English mastiff Morse (gentle, though as big as a horse), the athletic Dalmatian Bottomley Potts (covered in spots), the hairy and cuddlesome Old English sheepdog Muffin Mclay, (resembles a bundle of hay), the dachshund Schnitzel von Krumm (very low tum). And the mongrel Bitzer Maloney, all skinny and bony and "bits a' lots of dogs." Writing this sort of stuff looks easy, and it probably is.

Go! may be the only one worth living in. Who would not enjoy the original (though hardly inimitable) way the two dogs greet each other?

Poodle: Hello
Mutt: Hello
Poodle: Do you like my hat?
Mutt: I do not! [Pause] Good-by!
Poodle: Good-by.*

But the most remarkable dog book of all is the ripping, snarling, allegorical and bestselling *The Call of the Wild* by Jack London. London was born in California in 1876, and in his forty years he packed in a nomadic life of seafaring, gold-hunting, and general outdoorsmaneering; somehow he also found time for powerful storytelling. One early year in his biographical outline reads, "1891: Sails San Francisco bay on the *Razzle Dazzle*, carouses with waterfront roughs, engages in oyster piracy . . ."†

The Call of the Wild (1903) was his first smash, despite

* Eventually, after several attempts to gain the mutt's approval with a selection of hats, including one attempt on the ski slopes, the poodle wears a really elaborate hat that the mutt finally approves of, and they go off to a dog party together. There have, inevitably, been many philosophical interpretations of these exchanges, from the Lacanian to the Deleuzian, and the three things they all agree on is that this is not a feminist book, and the poodle is needy and the mutt is arrogant. Ludo has told me that he has done his best to "unlearn" some of these unattractive early imprints.

† If only all authors experienced this sort of character-forming early life before they picked up a pen. From *Jack London* by Kenneth J. Brandt (Northcote/British Council), which is strong both in the critical analysis of his novels and the attendant autobiographical context.

the apparently challenging subject matter—the singular story of a Saint Bernard–Scotch collie mix named Buck that starts as chronological narrative but ends as magic realism. Buck begins life as a reasonably mollycoddled pet in the home of a judge, but the judge has enemies and Buck gets kidnapped. He is then traded again several times, and ends up as a sled dog in the Yukon during the gold rush, finally with an owner who trusts and cares for him. Nothing exceptional here, but London's ultraconfident and visceral prose propels the reader to root not only for Buck but for all he represents: a bid for authenticity, a break from the shackles, a back-to-nature stampede. When his last owner dies, Buck makes one final transformation, a Darwinian reverse, leading a wild wolf pack into extreme adventure, ultimately ascending to become a primal and mythical Ghost Dog. Here he is, pursuing a snowshoe rabbit:

> He was sounding the deeps of his nature, and of the parts of his nature that were deeper than he, going back into the womb of Time. He was mastered by the sheer surging of life, the tidal wave of being, the perfect joy of each separate muscle, joint, and sinew in that it was everything that was not death, that it was aglow and rampant, expressing itself in movement, flying exultantly under the stars and over the face of dead matter that did not move.

Here was a dog as dogs used to be, London argues, before all the crossbreeding and pampering, and here was the dog truest to himself, triumphant in a hostile environment.

Hard to complain about the author's narrative vigor, much as one may question his argument. Were domestic animals truly happier when free? The fact that the book still works more than a century later is testament to its self-belief, if nothing else. Its follow-up, *White Fang*, is equally vivid, its subject this time closer to a wolf than a dog, with its loyalty purely to itself. Both books stun with their ruthlessness; if you read them as a pup you will never forget them, such is their expeditionary force.

But where is the most *memorable* dog in all literature to be found? Some would make a case for J. R. Ackerley's *My Dog Tulip*, an exacting, obsessively scatological and pungently autobiographical account of the joys of living with his German shepherd Queenie. Ackerley, a novelist and editor, came late to the charms of dogs, once insisting "that a firm stand should be made against British sentimentality over dogs—dirty, noisy creatures." Everything changed after falling for Queenie, who gave him the "incorruptible, uncritical devotion" he had desired all his life. "A dog has one aim in life," Ackerley latterly surmised, "to bestow his heart." And when she died he was beside himself, concluding "no human being has ever meant so much to me."

Others would stake a modern and assertive claim for the Great Dane in *The Friend* by Sigrid Nunez, in which the narrator inherits a dog named Apollo after the suicide of her mentor and lover. (Obviously a dog such as this will transform a life; Apollo enjoys being read to, particularly Karl Ove Knausgaard.) And then there are many supporters of Karenin, the mutt in *The Unbearable Lightness of Being* by Milan Kundera. Karenin's role is part baby substitute and part symbol of permanence in revolutionary times, and her death from cancer permits Kundera's narrator to ponder that a society may be judged on its treatment of animals; the suggestion is that we too often fail this test. At the end of his life, the main female character, Tereza, has a "sacrilegious" thought: "The love that tied her to Karenin was better than the love between her and [her husband] Tomas. Better, not bigger. . . . Given the nature of the human couple, the love of man and woman is a priori inferior to that which can exist (at least in the best instances) in the love between man and dog, that oddity of human history probably unplanned by the Creator."

I heartily recommend all of these dogs and their fictional, believable lives. But unfortunately I think the most memorable of all literary dogs is Bull's-eye in *Oliver Twist.* Why unfortunately? Because the poor dog is a chained monster, and his owner, Bill Sikes, is cruel to him at every turn. Dickens still hadn't owned a dog of his own by the time he wrote the novel in 1837, and he would later own and love many, but he clearly needed no personal experience to employ one as a literary cipher. When we first meet Bull's-eye, white and scruffy with cuts all over him, he is already accepting of his bullied fate. He is trained in viciousness, and his ineradicable loyalty to Sikes is awful to witness. When the dog leaps to his death after Sikes's hanging, it's a rhetorical demise as much as an actual one: he can't exist as the epitome of evil without his master.

The notion of dogs as menace was rife in Dickens's London. There was rabies (greater as fear than reality), and there was general canine wildness in the streets. The animals who avoided the wheels of motorcars may yet meet a worse fate at dog fights or a bear baiting. The creation of the Society for the Prevention of Cruelty to Animals in 1824 was partly in response to a backlash against dogs in this period.

And unruly dogs could inspire anyone to unruly thinking, not least an author trying to complete a novel. Sitting at his desk in Bloomsbury's Tavistock Square in December 1852, attempting to finish *Bleak House*, Dickens wrote to his colleague W. H. Wills that he was being "driven Mad by Dogs." A large group had "taken it into their accursed heads" to assemble every morning on a patch of ground opposite his window, and "barked this morning for *five hours without intermission*—positively rendering it impossible for me to work." But he had a solution. He would ask his manservant to see about the possibility of hiring a gun and some shot. "If I get those implements up here tonight, I'll be the death of some of them tomorrow morning."

But we need to end this literary voyage on something more pleasant than this, perhaps near where we began. In 1961 Vita Sackville-West published a collection of short and sweet character profiles of popular dogs, and like her own beloved salukis Edith and Zurcba they still sit well.* And no reason why they shouldn't: the character of dogs has changed little in the last sixty years; what

* Zurcba has also been referred to by her owner as Zurcha, which may suggest a certain insouciance. In the book, entitled *Faces*, Sackville-West describes Zurcba/Zurcha, who was given to her as a gift by Gertrude Bell, as "without exception the dullest dog I ever owned . . . she was completely spiritless, and as for fidelity she was faithful only to the best armchair."

has changed—through cognitive psychology and forensic science analysis—is our ability to understand why they have the character they do. Sackville-West chose forty-four breeds to accompany a portfolio of doleful and mischievous photographs by Laelia Goehr, and in her introduction to the collection she admits she has always preferred a large and noble breed over the delicate lapdog. She also apologizes to readers who may find her approach too anthropomorphic: "When one loves dogs, it is difficult not to attribute human qualities to them, so one almost automatically writes 'he' instead of 'it.'"

Owners of Labrador retrievers may balk at her description of the breed as a "faithful lump of a dog," but they will recognize her appreciation of their utter dependability and affection, and, as they age and thicken around the midriff, their mild disapproval of the antics of any younger generation. And what do they say? *Where's my dinner? Hope it's not been forgotten. Things aren't what they were.* Of the Kerry blue terrier she regrets that "his colouring is the only pretty thing about him. How would you like to be afflicted with such a beard, growing almost out of your eyes?" In the lamb-like Bedlington terrier she finds nothing playful, "a Northumbrian endowed with Northern hardiness . . . his wooden appearance does not much appeal to me, however linty his coat may be." And of the Pekingese, Sackville-West notes that despite appearances the dog is "no luxury knick-knack," even though she wouldn't want one for herself. The dog's eyes reminded her of a car's undipped headlamps.

Tastes differ, of course: if they didn't we'd just all own

Labradors. P. G. Wodehouse, for instance, believed the Pekingese to be "a different race and class," by which he meant vastly superior to anything else. "They may try to be democratic, but they don't really accept other dogs as their social equals."

Wodehouse had an obsession with his Pekes. He had many, and they had silly names and characters, and they seemed to sustain him when his writing and own company grew tedious. He realized that you don't get to spend a lot of time with one particular breed without picking up a few secrets, and seeing things the casual observer may miss. On January 20, 1936, for example, Wodehouse wrote to his friend and fellow author William Townsend. "Winks and Boo do nothing nowadays but fight," he began, before revealing the ability of Pekes to transmit "the Dirty Look." The Dirty Look was a version of the evil eye, and you had to know what you were looking for. "Winks and Boo will be sleeping quite happily at different ends of the room," the writer explains, "and then suddenly one of them will lift her head and stare. The other then stares. This goes on for about ten seconds, and then they rush at one another." What is the cause of this fracas? Wodehouse thought he knew: "Dogs say things which the human ear can't hear."*

---

* You liked that last shaggy dog story? How about this one, possibly from Mexico. Since the world began, we have been able to divide its inhabitants into two groups: those with a warm generosity of soul and those without. Those with goodness in their hearts look after creatures less able to look after themselves, but those with a harsh temperament willingly administer cruel blows in the hope of self-advancement. Dogs have been on the receiving end of both gratitude and cruelty. So what can be done to rid us of the latter and make this a better world for us all? Many dogs have long

Is there a dog lover in the world who would not concur? Is there anyone who doesn't believe—with absolutely no scientific evidence to back this up—that their dog is not only the most intelligent and gifted to have ever peed on a tree, but has powers as yet unexplained by the rational? And why stop at dogs who merely say things? As the following chapter makes clear, humans have been making the most of the learned and performing dog for centuries, although not all performances have been honorable.

---

believed in divine guidance—God, perhaps, or a benevolent alignment of the stars. And to this end in ancient times did they ask their higher spirit for assistance. So when a single dog passed on and ascended to the next world, she was accompanied by a written message for the higher spirit; the message was placed just inside her bottom for security. It read: "What may be done to rid the world of beating and harsh attitudes?" Alas, we are still waiting for the reply, but hope lives eternal. And that is why, every time dogs greet each other, they examine each other's backsides for the return message: perhaps one day the answer will come. (I love this story as much as the next dog, but the reality is that dogs sniff each other's behinds because that's where all the personal information is stored. It's like a handshake, but more revealing: the hormones excreted by the glands surrounding the genitalia tell dogs about gender, age, diet, health, temperament and whether they've met before.)

10.

Through the Hoops

What do we want from our dogs? Why do we want them to be more than they are? One answer is simple: we want them to be more like us, to bolster our bond in ever more ludicrous and highly amusing ways. We will not let a dog just be a dog, not when they can also be a performing dog.

For at least two hundred years we have been asking ourselves whether we would like to see a dog play dominoes and cards. Or whether we would benefit from a dog dressed as Carmen Miranda serenading another dog wearing a sombrero. Never mind our ability to read a dog's mind in an effort to learn what they know and what they need—how about a gussied-up show-business dog that can purportedly read *our* mind?

In September 2019, with this book nearing completion, I took part in the Milton Keynes Literary Festival. It was a nice lineup—Jack Monroe, Carrie Gracie, Paul Mason, Lissa Evans—and I was happy to join them to talk about my work in progress. I discussed some of the content you've just read, and the audience was politely appreciative, but what they really liked was a slightly surreal set of conversations I'd concocted between my Labrador, Ludo, and other dogs he met on Hampstead Heath. My thinking was that only trivial dogs would have trivial conversations; I wanted to propose that dogs knew far more about us and the world around them than we often gave them credit for. One conversation, on a bright spring day around the back of the Parliament Hill athletics track, involved Ludo and a rough collie named Aperol.

Ludo: Hello.

Aperol: Hello.

Ludo: Have you ever taken an IQ test?

Aperol: Yes.

L: And?

A: And it went wrong. There was a snack and a beverage for the examiner that I assumed was for me, so a bad start.

L: You ate the examiner's snack?

A: And his beverage.

L: Oh. How did the actual test go?

A: I don't want to talk about it.

L: Understood. Ha ha!

A: Ha ha!

L: What beverage was it?

A: I can't remember. Oxtail.

L: Oh. Yesterday I spent two hours in the televised company of Cesar Millan, the Dog Whisperer.

A: I hate it when they whisper. It makes me frightened that suddenly they're going to shout something very loudly. Also, it can destroy the ears, and no amount of linctus will fix them.

L: Cesar was whispering to a very, very angry Doberman who attacked visitors' legs. Cesar kept on saying, "Shh-hh!" and he managed to change the Doberman's preferences. In the course of two weeks he weaned him off legs and on to furniture. The owner was furious!

A: Ha ha! *Chippendale furniture!*

L: Ha ha! All our doors are made by Chippendale.

A: Chippendale is just another name for wood.

L: Have you heard of Ogden Ganache?

A: Yes!

L: Really?

A: No.

L: Ogden Ganache said that the definition of a door was something that dogs always wanted to be on the other side of.

A: They open the door if you whimper. Yesterday someone came for tea who once knew Rachel Carson.

L: The woman who wrote the book *Silent Spring* in 1962 that set off a debate about how we were destroying our planet by the use of pesticides, a brave and brilliant clarion call that inspired an environmental movement to rise up to try to reverse this terrible damage?

A: Yes.

L: Amazing. She was *very* influential. But of course no one listened to her. And now we face disaster. I notice it just on a gentle walk.

A: So do I.

L: The grass is not the same. The mud smells different. There seems to be less of everything around.

A: We notice it.

L: Us dogs on Hampstead Heath are another early warning.

Of course this was a totally anthropomorphic act: Why should I assume that dogs speak English and employ a recognizable syntax? Why should Rachel Carson be a recognizable figure in the canine world? (In other conversations, Ludo and his friends discuss Jeremy Irons, the Kinks, George Orwell, Motörhead and the Netflix drama *Chernobyl*.) Partly this is because we know no better; in the absence of an alternative, unified dog language, we are only able to imprint our own. (As discussed previously, we interpret barking, growling and emissions of air from the nostrils as a subtle sequence of alerts; each indicates a different need or emotion at different times, according to circumstance, dog and our own ears.)

But such a conversation is also a loving act: dog owners have always found it hard not to conjure up an interior life for their charges, and of course we speak to them as if they understand. It only requires a small logical leap to imagine dogs talking back, and to each other. When Shakespeare puts a dog named Crab on the stage in *The Two Gentlemen of Verona*, he is accused by his melodramatic owner Launce of being "the sourest-natured dog that ever lived," and is berated for expressing neither tears nor words when an emotional situation demanded it. There was clearly an expectation that Crab *would* speak.

These days, the talking dog has become a familiar schtick, and we may only ask what took it so long. The talking animal began with the serpent, was popular at the time of Aesop, middled out with *Alice*, and reached maturity with *The Jungle Book*, *The Magic Roundabout*, *Toy Story* and *Zootopia*. (And even mute animals can speak to those who can hear: "What's that, Lassie, old Maggie Finnegan has fallen down the disused well again?") No one makes a fuss about talking dogs anymore, and the acceptance extends well beyond Disneyfication.

In the autumn of 2019, a short story in the *New Yorker* entitled "The Fellow" by Joy Williams featured a dog who expressly wished his coat color to be described as "Devil's-food cake." The dog had a melancholy air, for the dog had seen most things before. The dog seemed to be able to look inside the narrator's soul, and showed an interest in the poetry of Craig Arnold, while also maintaining a reliable line in sniffing and chewing. Toward the end, the dog reveals that he or she has died many times, and may have come to represent all dogs. Two further observations from the dog: he or she is "not a fan of electronic devices," and regrets "being taken from our home and expected to thrive in some other place."

At Milton Keynes, Ludo did thrive, of course. In fact, he stole the show. What the audience liked just as much as Ludo's conversations was the fact that I'd brought Ludo along with me. He was with me onstage, sometimes in his bed and sometimes adjacent to it, twelve years old now but still able to command a room. I could have been spouting any old nonsense up there—any old stuff about

Buck, Tulip or Bull's-eye—and the crowd wouldn't have minded; but occasionally his inquisitive mind wandered to the people in the front row. Perhaps they had a bit of food in their bags; perhaps some of it would be for him. So he hopped offstage and wandered about, the audience loving every minute of his impromptu visits along the aisles, even when they caught a whiff of his fish-breath. When I called him back to his bed next to me, he only managed to get his forepaws onto the stage, requiring that I climb down to shove him up. Once up he upstaged me again, this time by yawning when I was telling the story of early domestication.

And so by popular demand, and with the understanding that it may be his positively last appearance in the town, for who knew when this velveteen twelve-year-old would once again summon the strength or enthusiasm to tread the boards at this unusually cultured junction off the M1, Ludo appeared once more "in conversation." This time the scene is early one spring morning on the edge of the male bathing pond with Milo the lurcher.

Ludo: Hello.

Milo: Hello.

Ludo: You can't possibly guess whom I met yesterday.

Milo: John the boxmas.

L: Correct! How did you know?

M: Because I met him too. He's a boxer-mastiff blend. There are only two boxmases on the Heath, but one is on holiday. They called him John to instill a sense of proportion and normality. His temperament could be best

Just like us? A Victorian gentleman prepares for a big night.

The telltale image: a Canaan dog at Sha'ar Hagai Kennels in Israel . . .

. . . and an early ancestor on rocks in the Sahara desert, Algeria.

Beware of the dog: a mosaic warns intruders in Pompeii.

Ancient dog seeks new owner: a bronze totem.

Bulldogs stick together: *A Friend in Need* by Cassius Marcellus Coolidge.

A wet William Wegman Weimaraner mosaic in a New York City subway station.

An Elliott Erwitt classic in New York City.

One of the first great dogs of literature: *Flush*, Virginia Woolf's novelized "biography" of Elizabeth Barrett Browning's cocker spaniel.

Charles Dickens and his dog.

The novelist J. R. Ackerley with his German shepherd, Queenie, of whom he wrote, "No human being has ever meant so much to me."

Tales of joy in *The Tail-Wagger Magazine*.

The ascent of dog in 1872: Charles Darwin detects submission, fear, and devotion.

Darwin also suggested that dogs have a sense of humor. It's hard to argue with him.

A puppy less than impressed by *The Origin of Species.*

Today, we've gained insight into how the smartest dogs learn. Chaser, shown here with her toys, learned the names of more than one thousand objects.

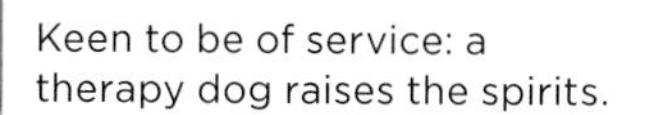

Keen to be of service: a therapy dog raises the spirits.

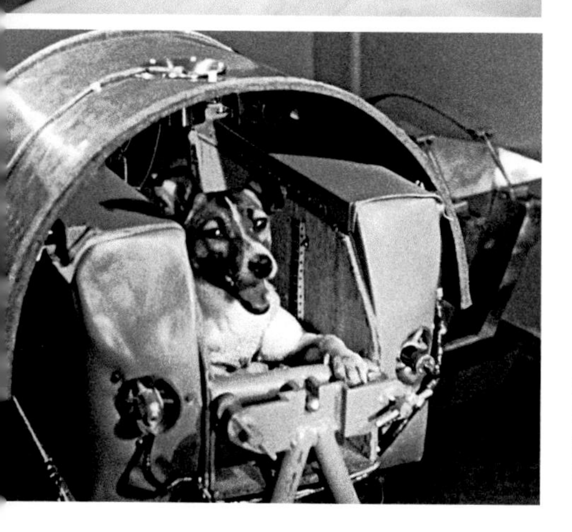

Dogs were also sent to the farthest frontier. Here, Laika prepares for space. Her mission did not end well.

Charles Cruft, the first great dog showman and the namesake of the British dog extravaganza Crufts.

Una Troubridge and Radclyffe Hall with their dachshunds at Crufts in 1923.

Ready for the ring: a proud English springer spaniel and an even prouder handler at Crufts.

The dog is the star: Mick the Miller films *Wild Boy* in 1933.

Snoopy finally gets his own Tokyo museum in 2016.

The most famous corgis in the world: Princess Elizabeth holds Jane while Dookie stands guard, 1936.

A labradoodle, the original designer dog.

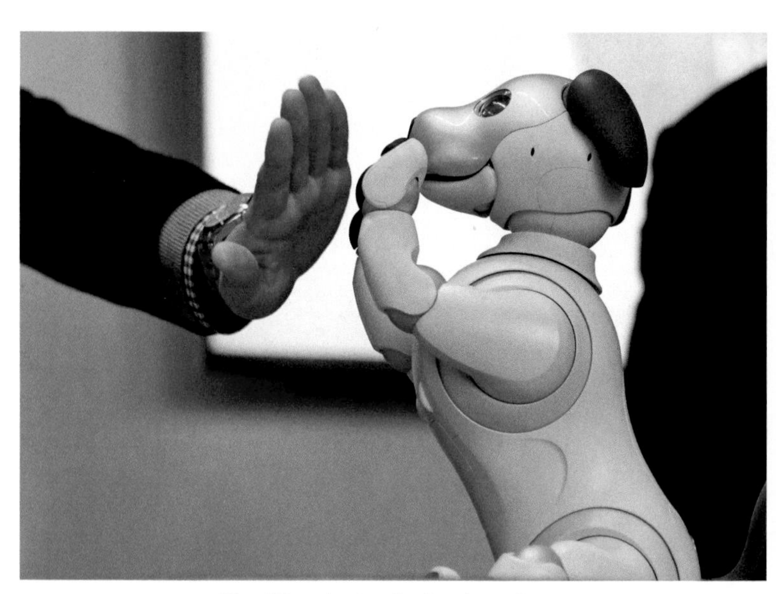

The Aibo electronic dog from Sony.

All good dogs: the Victorian dog cemetery in Hyde Park.

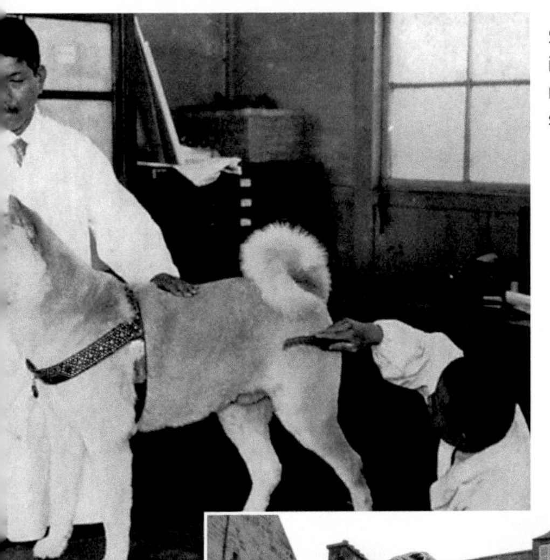

Some dogs are destined for immortality: here, Hachikō, a national hero in Japan, gets stuffed in 1935.

Edinburgh's Greyfriars Bobby, forever standing watch.

The author’s beloved Ludo, as a puppy . . .

. . . and as a distinguished older dog.

described as unmanageably loose, but he's only loose when *he* wants, not when *you* want.

L: That's typical of the boxmas.

M: He's scary to look at but nice when you get to know him. Children scream when he approaches, but they soon dry their tears.

L: Ha ha!

M: Ha ha!

L: Do you like mild adult themes combined with a vague parental unease?

M: Always.

L: Because last night they all watched *Family Guy*.

M: The one where Brian the talking dog becomes the celebrity mascot for a beer company and becomes so popular that they make a huge balloon of him for a fancy parade, and then Peter gets jealous and tries to shoot down the balloon with a crossbow?

L: No.

M: The one where he dies in a car accident and causes such distress among viewers that he's brought back a couple of episodes later with the help of a time machine?

L: No, it was another one. But Brian is the best type of talking dog because he is so snarky that you soon forget he is unusual and regard him as just another character in the animation. You look forward to seeing him.

M: *Talking dog*, ha ha!

L: Ha ha!

M: They've just launched the new AIBO.

L: Amazing!

M: You don't know what it is, do you?

L: No.

M: It's the latest electronic dog. From Sony. He's Version 2.

L: Is it plastic and full of electronics?

M: Unfortunately, yes, but it's the most domesticated model you can get, apart from a real dog. The instructions tell humans how to operate it, and the more you play with it, the more it learns.

L: Like a real dog.

M: Yes. It's very primitive artificial intelligence, or AI, or AIBO. The most intelligent thing about it is that it doesn't get skin infections.

L: Does it wee?

M: It's advanced, but not that advanced. It lifts a leg sometimes, but nothing comes out, so in that respect he's a bit like you. Perhaps he'll wee in Version 3, which may also come with a heart monitor. He's a cute scamp with a waggly tail, like a village dog, and the instructional video shows grown adults stroking his back and tickling his chin, even though he's plastic.

L: How can you tell it's a he?

M: Because he gets stuck in corners and won't turn around. Version 3 will be a she and be able to turn around. Version 2 can sing and pick up a plastic bone and recharge overnight. He has a camera in his nose and he also has a microphone, and he sends back data to Sony.

L: The humans are gaining even more control over us. They should call him Snoopy. Will Version 3 be able to hook up with Version 2?

M: Yes. It will be like *Gremlins*, and soon there will be

AIBOs everywhere. Each one costs $2,900 and apparently they're very popular among idiots.

L: I want a snack.

M: I want a snack too.

[Distant voices: Milo? Ludo?]

L: Ignore them. They never leave without us.

M: My owners smell of sweet corn, coffee, earth and floof.

L: You said "owners." Non-PC!

M: I know, but what else have you got?

L: I've always liked "grown-up feeder and walking device."

M: Ha ha, *grown-up*!

L: Ha ha!

M: My owners smell of banana bread, floof, ginger and Lenor. The woman over the road smells of Comfort.

[Insistent: Ludo!? Milo?!]

L: Well, goodbye.

M: Yes, goodbye, Ludo. *À bientôt.*

Talking dogs? But of course. We have grown used to them from fiction and the movies, and from the performing dog of the Victorian parlor—a dog playing canasta and dominoes, a dog walking on its front legs straight from the Folies Bergères. We have become accustomed to all manner of entertainments in which a dog becomes more like a human.

The story is told of Munito, a large white mutt with a brown patch over his left eye, the so-called Isaac Newton of his species. There wasn't much Munito couldn't do, though he drew the line at explaining how he did it. According to Ricky Jay, the late chronicler of bizarre feats and hucksterism, Munito was schooled in botany, geography, natural history and palmistry, and was able to solve puzzles of addition, subtraction and multiplication by selecting an appropriate card from many choices in a large circle. (On a commemorative china plate, which shows him playing dominoes, he looks like an elaborately semi-shorn poodle, and though his breed is never specified, he was believed to be a cross between a Scottish deerhound and an Irish water spaniel.) In 1817 Munito appeared twice daily at Laxton's Room at 23 New Bond Street in the proud wardship of a certain Signor Castelli, who charged a shilling (Castelli was a snake-oil kind of name; in reality he may have been a Dutch trainer called Nief). Castelli/Nief and Munito achieved even greater fame when he performed the more traditional dog duty of rescuing a drowning woman from a lake in Green Park (for which Munito received a bravery medal, previously only awarded to humans, from the Royal Humane Society). Although here too all may not be as it seemed: the rescue may well have been a publicity stunt, the drowning woman happily "drowning" on cue.

After success in England, Munito performed shows in Austria and Finland. Ricky Jay notes that when he returned to London, advertisements announced that his sojourn abroad enabled him to "finish his education." Munito's fans included the Prince Regent and the Duke of York, but none seemed more impressed than Charles Dickens. Many years later, in 1867, Dickens recalled in his journal *All the*

Year Round how he witnessed the dog's extraordinary performance twice when he was a young man. At the second show he detected how most of his tricks were achieved: Castelli would apply a trace of aniseed oil to a card with his thumb, and Munito would sniff it out. There were other explanations, including E. de Tarade's theory advanced in *Education du Chien* one year before Dickens. Here, sound was the key; when Munito approached the correct card, Castelli would click his fingernail in his pocket, and the dog would pounce.

Munito's notoriety attracted many imitators, among them a dog named Monetto, who could tell the time. Performing dogs were also a common sight on the legitimate stage. In the mid-eighteenth century the Animal Comedians appeared at the New Theatre, Haymarket, in a battle scene enacted by sword-wielding dogs and monkeys that ended with the performers "singing" "God Save the King." Some fifty years later *The Caravan* at the Theatre Royal, Drury Lane, featured a scene familiar from oil painting—a Newfoundland named Carlo leaping to rescue a boy drowning in a water tank.

Carlo became so popular that in 1809 he wrote his autobiography. His book was placed upon a crowded shelf, for dog memoirs were all the rage. We have already encountered the spotted terrier named Bob who published *The Dog of Knowledge*, a volume of false modesty if ever there was. "I shall neither use disguise nor concealment," Bob begins, "I shall neither boast of exploits that I never performed, nor seek to build my reputation on the ruin of another's name." Not all humans shared such equitable ambitions, Bob feared, for it was a dog-eat-dog world out there at the dawn of the nineteenth century. "Happy would it be for mankind if the two-legged puppies . . . were equally candid and forbearing. Instead of trying to snatch the bone from their neighbour's mouth, or snarling when they happen to meet, it would be well if the interchanges of humanity more frequently took place, and the strong and powerful lent their ready aid to the helpless and the weak."

Philanthropy was only one of a dog's unanticipated talents. In the 1890s a French poodle named the Inimitable Dick (and his many imitators) reproduced the glamorous Serpentine Dance made famous by Loïe Fuller, an act that demanded the attachment of stilts before Dick twirled within layers of diaphanous blue silk. The same decade saw the cacophonous exploits of Mr. Louis Lavater's Dog Orchestra, an array of six mongrels in fancy garb. A photo in the *Strand* magazine from 1897 shows a dog called Jack astride a little stool playing the trombone. The article described the orchestra's ignominious formation in Holland: one member had a stick tied to his paw so he could beat a tea tray, while another crashed into his bass drum and Trombone Jack overbalanced and fell into the pit occupied by the human orchestra.

By popular demand: a positively last appearance. The performing dog—like so much that marked the dog as a product and concept rather than just an animal—came into its own in Victorian times. But as we view these acts with a stern countenance from our lofty moral pedestal of today, we should note that even the Victorians had a sense of humor about their entertainments. Boswell the Unusual

Eater of Bern, for instance, was not only a dog who ate pencils, sandals, keys (single and bunches) and towels (hand and bath), and not only a dog whose enormous appetite surpassed even that of his cousin, the Noble Walter, but also (according to his promotional handbill) a dog whose "personal demeanor" has been "known to melt the frostiest embankments of the Swiss countryside." He was also, *mais oui*, a connoisseur of the arts: "his literary pretensions are too well known to be ignored."

Perhaps we shouldn't be surprised to learn that in 1918 Harry Houdini owned a terrier named Bobby who performed an escape from manacles and a straitjacket (nor that Bobby had a rival in Paris, another Newfoundland, who answered to the name Emile, nor that Emile had a rival of his own, known as Nelson, who could also spring himself free of restraints with a single bound).

Why are we so keen to encourage such human antics from our closest animal friends? And why are we happy to be semi-fooled by them? With regard to the nineteenth century, one answer is bound up in the mania for spiritualism, a belief and investment in a life force beyond our own. Those who worshiped Egyptian gods with the heads of dogs would have recognized this trait. Another answer lies in the value we place in our dogs as entertainers. For all the historical weight attached to the traditional bond between human and animal, surely the role of fun is underrated. The base inquisitiveness of dogs is often naturally amusing to us, and we find no difficulty in amplifying this with the addition of human curiosity. How would it look if we wrapped goggles around a poodle's face and put her behind the wheel? Why not place a paw on a typewriter and wait for the likes? The question, always, is how far to push it.

IN SEPTEMBER 1992 as a journalist I flew to Las Vegas to report on the terrible aftermath of a dog act that everyone I spoke to said was the best in the world. Gerard Soules was the uncrowned Poodle King, a fifty-five-year-old man who ran "Les Poodles de Paree," a show so sophisticated that it was easy, amid the seductive lighting, to imagine it really was a tipsy, smaller Carmen Miranda or Marilyn Monroe up there onstage. But it did not end well.

The ten-minute act was watched with open jaws, for here were poodles like you'd never seen them. A dozen creatures on their hind legs, camped up in feather boas and hats and little sequined French gowns, hula skirts and sombreros. One of them was modeled on Mae West, skipping and shimmying to "Mimi," "Thank Heaven for Little Girls" and "The Mexican Hat Dance." To the accompaniment of the "Wedding March," one poodle was dressed as the nervous groom at the altar while another, the bride, waltzed up the aisle to join him. And then there was the grand finale, in which all twelve dogs high-kicked their way around the ring to the cancan. "Once seen, never forgotten!" the ringmaster barked, with some authority.

"It actually stopped the show," said David Cousans, an ice-skating juggler who saw Soules's act as part of the Ice Capades show in Mexico. "Gerard was on skates, the dogs were on a mat, and he used to skate around and pick them up at the other side of the mat. It was very funny, very pretty. He made all the costumes himself."

Soules worked Radio City Music Hall, all the big ice shows, and most of the famous cruises. "He was *there*," declared Kenny Dodd, a former professional clown and a friend for many years. "He had a very good salary. There may have been acts that earned more, but none worked so hard. Maybe they worked twenty weeks a year, whereas Gerard worked fifty weeks a year. He could make maybe fifteen hundred dollars a week in Las Vegas, maybe a thousand in an amusement park."

But the dog act was not his first. Soules was born and brought up in Livonia, Michigan, and had loved show business as a child, particularly the high-wire and trapeze artists. "It was like in the fairy tales. When he was five he packed a bag and told his mother he was off to join the circus. He was back home within two hours," his sister, Colleen Anderson, told me. Anderson, two years Gerard Soules's junior, remembers him as "the best brother anyone could ever hope to have."

Soules was particularly entranced by home movies he'd seen of a woman named Betty Ruth doing a risky "heel-catching" routine on a trapeze, and he resolved to emulate her. He gained a reputation as a driven, fearless performer, admired by other acts. "Be brave but never reckless," he told his colleagues. He was also archly flamboyant: he flashed a cape on his entrance and exit, and he preened unashamedly. His friend Max Butler, who ran a magic shop in London, told me that his attitude toward his audience was, "If you don't like me, you're crazy." Butler used to tell people, "Oh, Gerry's great—and if you don't believe it, just ask him." The European promotional material for a Ringling Brothers and Barnum & Bailey show in which he appeared in 1963 said of Soules that "by his supreme artistry he soars to new heights of aerial audacity."

Soules didn't use a safety net, and invariably there were falls. One day in 1964 Gerard Soules just quit cold. He called it his *crise de nerfs*: he explained that all he could see from his trapeze was the ground below. But some years before, as he was recovering in the hospital after an accident, a friend had bought him a puppy. "You know," he told his friend, "I've always wanted to have a dog act. But I won't be cruel and teach them jumping tricks. I'll do something special."

Soules's poodle show took about a year to piece together. The idea for the fancy dress evolved from Soules's love of the musical; he believed that what people really wanted from their entertainers was not so much danger as glamour. "I start training them when they're a year to fourteen months, using the 'pork chop' method," he explained in a press release. "By that I mean plenty of rewards like love. Training sessions are more like playing with the dogs than working with them."

Soules's methods were a great improvement on those reported a century before. In a letter to the *Observer* in November 1913, a certain Basil Tozer reported having seen several performing dog acts over the years, many the products of gentle trainers who treated their animals with respect. Visiting the training sessions after one show in

Paris, for example, featuring dogs and birds trained by a man named McCart, Tozer saw only kindness, while admitting that the animals in question did not "do anything very wonderful."

Wonder, alas, required a level of pain and a fear of reprimand. Tozer went undercover to learn the training methods of a Continental man he referred to as Z, whose popular show in London featured dogs walking on their forepaws with their hind legs in the air. "He hangs the dog by its hind legs to a bar, and leaves it thus suspended." The legs were tied together with a strap, the strap attached to a hook. "When the dog is thoroughly exhausted Z returns to it, speaks to it kindly, and then places his hand so that the dog's forepaws can rest upon the open palm, the strain upon the hind legs thus being temporarily relieved." This system is repeated, the dog once again being hung from a bar in pain, until it comes to understand what is required of it. "If the dog fails to stand up and balance itself . . . it is given several sharp cuts across the belly."

Basil Tozer described another trick in which one of Z's dogs tumbles over on its side "like a drunken man—a trick which never fails to evoke roars of laughter." How is this achieved? By training the dog with a belt of pins upon its stomach, and a punishment involving pellets from an air rifle. Fellow entertainers who witnessed such methods were predictably outraged; Tozer described how, backstage at the music hall, an angel in tights and wings performed an angelic and avenging act by hitting a dog trainer across the mouth with the back of her hand.

By 1992, Gerard Soules's "Poodles de Paree" were parading five times a day in the big top at Circus Circus Hotel in Las Vegas. He lived with his dogs in a trailer park some miles away, and he had hired a man named Fred Steese to help him with transport and handling. Steese became Soules's lover. But he also had a criminal record, and when he had difficulty gaining his official work documents from the hotel, Soules had to let him go. It was not a happy parting.

Sometime later, in the early hours of June 4, 1992, Soules was found murdered in his trailer from multiple stab wounds, his distressed but unharmed dogs squealing nearby. There were no obvious suspects, but when the police picked Steese up hundreds of miles away and found Soules's name among his contacts, they charged him with his death. While always maintaining his innocence (claiming he was several states away at the time of the murder), Steese was convicted in 1995 and sentenced to life imprisonment. The conviction was overturned in 2013, and Steese went free with a full pardon. Toward the end of 2019, with Steese applying for compensation, Soules's murder remained unsolved.*

After his death, Soules's poodles found a new home at the small Circus Corona in the Midwest, although they too have long since passed on to the never-ending costumed parade in the sky. Those who saw them perform on earth still claim their show could never be bettered.

---

* Recalling his funeral, his sister Colleen Anderson told me, "After the service at the graveside my brother Jim stood up and said, 'Gerry lived for applause—let's give him one final send-off.' So we all stood up and applauded for ten minutes."

• • •

AND THAT'S what they once said about a dog called Mick the Miller.

Like poodle acts and pigeon fancying, greyhound racing has fallen victim to the modern age. Most dog tracks have closed, the dog men retired. But not so long ago these dogs were famous and everywhere, performers who turned hard men soft and soft men hard, and made a few of them rich but most of them poor.

The sport, such as it is, limps on in a few British locales, but largely it has gone to the dogs.* West Ham greyhound track is no more, and neither is Catford, which did its best among the new money of the 1980s and the stag parties and the ironic postmodern marketing but eventually had to admit its time had come. Television companies failed to find ways to make the track exciting, for the five-hundred-yard races were over in a blur, and when animal welfare organizations exposed the cruelty faced by greyhounds during their life on the track, and appealed for dogs to be rehoused after it, the spectacle lost its allure. Large parts of Australia and the United States have banned the sport outright.

Partly, dog racing diminished for the same reason as

* The origin of the phrase is uncertain, though it always tends to denote a decline in personal well-being and circumstance. It certainly gained popularity at the dog track, where the races created many paupers. It may simply imply something rotten and on the downward turn—food no longer fit for human consumption would go to the dogs. Or it may have originated outside ancient city walls in China, where criminals and other undesirables would join similarly shunned creatures, including dogs.

British wrestling: the stories dried up; the sport ran out of heroes; it became socially unfashionable. And perhaps in the end we just loved our dogs too much to treat them this way. The bond between man and dog at the track—fragile in so many ways (the stupidity of the concept, the ill-treatment of beautiful animals, the hoodwinked venality of the punters)—could only be maintained while the romance held its luster. These days we may have trouble recalling just how strong that romance was, and hardly anyone beyond habitués of the few tracks that remain is able to identify even one greyhound by name.

But once there was a dog known by those with no interest in greyhound racing, an animal that transcended the sport. When people meet today to draw up lists of the Top 100 Sporting Heroes of All Time, that dog always makes the cut.

Mick the Miller is preserved in a glass case in an outpost of the Natural History Museum in Tring, Hertfordshire. He is not the most handsome of hounds, and certainly not the most striking in the museum; the stuffed Great Dane next to him looks as if it could eat him whole. But the stuffed Great Dane garners little passing affection from visitors, while the sharp brown greyhound absorbs all the love in the room. Somehow, almost ninety years after his last race, the dog still looks ready to go, and to win. How can this possibly be?

"Mick had that strange quality of separateness which made him look as though he, and he alone, could do the things that he did," writes Laura Thompson, who grew up with racing dogs. "At the same time, when he did these things, he took the dog men with him. They liked him because he looked like one of them: rather plain, rather ordinary, low and rangy about the head, compact and finite about the body, invulnerable and streetwise about the eye."

Mick the Miller was a working-class hero in a time of national malaise, and his astonishing acceleration suggested escape. (And all this foisted upon a breed that once epitomized the most aristocratic of thoroughbreds; Henry VII featured a greyhound on his heraldic shield, and Henry VIII could find no finer company for the hunt.) Everyone who saw Mick in his prime—the seventy thousand who turned out one scented summer night at White City in June 1931, for instance—would live for years in his fabled slipstream. During the depression, following Mick the Miller was like following a rocket to the moon.

Modern greyhound racing—several dogs snapped from their traps to pursue a mechanical hare for a few hundred yards around an oval track, and always failing—was an extension of hare and rabbit coursing from the eighteenth century, a blood sport turned pure sport. Initially the races only featured two dogs at a time. The business began in Manchester in 1926, just four weeks after Mick the Miller's birth, and soon afterward there was hardly a patch of wasteland in the country that wasn't being converted into a track. At its peak in the 1940s there were upward of 70 million visits to races at 77 officially governed tracks and more than 250 unlicensed ones. Today, with attendance below 2 million, 25 tracks remain; London, which alone once had 25 venues, now has none.*

The greyhound accelerates faster than any mammal save the cheetah, with peak performance usually reached in their second year. But for Mick the Miller speed was only a part of it: he was also a brilliant negotiator on the track, finding gaps to tear through to stay out of trouble. The dog just seemed to have the natural brain for it, and when he raced he appeared to be exuberantly enjoying himself; looking back, no one could ever remember him snarling, or remember his eyes popping in fever from his face like the other dogs'. As a performer he was Fred Astaire.

His achievements on the track were unmatched. He won his first big race at the age of three in 1929, the classic Derby at White City. He won the Derby again the following year, one of only two dogs ever to claim consecutive trophies. He was the first dog to win nineteen consecutive races, a record unequaled for more than forty years. In the space of three years, which is the regular span for a greyhound, he won fifty-one of his sixty-eight races, failing to finish in the top two only five times. He broke six world records at various distances, and his twenty-four trophies brought in more than £9,000 in prize money. He was the only dog in history to win all three classics: the Derby, the St. Leger and the Cesarewitch. Reporting on his second Derby, the *Greyhound*

---

* Greyhound racing does continue to attract considerable betting interest through bookies and online, if not in person. The legalization of off-track betting was another factor in the decrease in attendance. The dogs' role as objects of investment increased in direct proportion to their diminished role as animals with character.

Evening Mirror (which, at the peak of the sport, was a daily publication) observed how Mick's acceleration at the first bend, and his "amazingly clever" way of hugging the rails, "evoked yells of admiration, and as the favourite passed the winning line three lengths in front of his nearest rival Bradshaw Fold, the ecstasy of the crowd developed into a terrific crescendo of cheers."

The glory did not end with disappointment, as with so many sporting lives, but with triumph: Mick the Miller won his last race. And the glory continued when he hung up his vest. His trainer Sidney Orton put him out to stud at fifty guineas a pop, earning his stud master, Jack Masters, somewhere in the region of £20,000—enough money to buy a grand house and a racing car.

Mick the Miller died four months before the outbreak of the Second World War; it was as if the two things couldn't possibly coexist in the same constellation. His funeral would have drawn thousands, the ultimate flat cap parade. But the Natural History Museum didn't have many contemporary stars in its cases, and it swiftly stamped a preservation order on him. When his body was prepared for taxidermy, his heart weighed almost two ounces more than the average.

Today, proud against a fawn backcloth, his keen eyes are glass. He wears what visitors take to be a faint smile, and his unusually long tail, with the familiar white tick at the end of it, still draws comments about aerodynamic balance, the way a sports car uses a spoiler. His present devotees are still looking for new ways to remember him, his achievements as a sprinter long ago supplanted by his manifest representation of a dream. In 2011, some four hundred people, the Taoiseach Brian Cowen among them, turned out in Killeigh, the village of his birth in County Offaly, Ireland, to unveil a sleek, life-size bronze statue on the village green, where he looks like a valiant war hero.

Mick the Miller united a nation like no other dog. When Mick ran, it was said that even the owners of competing greyhounds wanted him to win. "Dog racing loves life," Laura Thompson declares. "It is a celebration of life forces; expectation, hope, desire, greed." In the middle of the last century, humans could find no more suitable animal to represent these emotions than their dogs. Between 1930 and 1980, despite all its deviancy and exploitation, dog racing proved itself—and perhaps never quite as much as when Mick was making mincemeat of the bends—as another embodiment of our enduring bond. But the fact that it wasn't of mutual benefit hastened its demise.

THESE DAYS we find other ways to express our admiration for the performing dog. Dogs continue to delight on blooper and talent shows, and for years, "Stupid Pet Tricks" was one of the most popular segments on *Late Night with David Letterman*: dogs attacked vacuum cleaners, delivered beer, climbed ladders, rode skateboards and played basketball.

The most enjoyable pet tricks on YouTube are often the ones with a) the least apparent training of the dog in question and the most spontaneous nuttiness, and b) the ones that entail the dogs displaying the most human-like traits. If you have never seen the extraordinary performance of a dog called Clark Griswold, a rescue Dutch shepherd/Malinois mix from Colorado, then where have you been? The video—google "Ultimate Dog Tease"—has been watched on YouTube about 200 million times, eliciting such comments as "That dog looks like he is actually talking," and "I really wish dogs could talk." Another viewer has observed, with some justification, "What a great time in history."

Clark is a disappointed dog. His owner teases him with the mouthwatering possibilities of meals in the fridge—not least the multilayered potential of bacon—and Clark responds attentively with "The maple kind, yeah?" His owner teases, "I know who would like that!" and Clark is hopeful. But then his owner says, "Me. So I ate it!" Clark then looks away and lets out a huge doggo yawning sigh—"Awwwwww!"—followed by a less dramatic and sadly resigned "No, you're kidding me!" The game resumes. There is also chicken and cheese in the fridge, and before the owner says he covered this combination with . . . he is interrupted by Clark asking, "Covered it with what?" Alas, the treat all goes to the cat, and on hearing this Clark lets out a huge and exasperated howl. The voices are synched superbly, and the situation entirely believable. Every dog would love those meals, and every dog would respond with a similar sentiment when they were so cruelly snatched away.

But that's not actually Clark's owner we hear. The video is the work of Andrew Grantham from Halifax, Nova Scotia, who runs a YouTube channel and merchandise outlet called Talking Animals from his home (T-shirts and mugs are emblazoned with such phrases as "Covered It With What?"). Clark's real owner sent him footage of an inquisitive Clark in the kitchen, and Grantham supplied the vocals. (Grantham's other efforts include a talking cat called Jupiter and various hamsters, mice and fish.) One may, of course, object to its silliness and its shameless anthropomorphic exploitation. You could accuse Grantham and his imitators of amplifying our familiar treatment of dogs as human playthings, as mere extensions of ourselves. But what sort of miserable soul would do that? To deny the sheer absurdity of it, and the unabashed adorability of it, would, I think, deny us one of the most enjoyable social interactions we have with our dogs. We talk to them every time we engage with them, even if not always with words. Admittedly the human gets more out of this particular transaction than the dog, but Grantham's merchandise sales benefit a dog charity, and if even the smallest percentage of Clark's viewers are tempted to visit a rescue center, then good for them and the dogs.

Inevitably, Clark the dog has his own Facebook page, on which he lists his personal interests (pretty much the same interests as every other dog in the world but with double bacon), and he also lists his favorite poem, a slinky bit of corn by the Californian dog trainer Janine Allen, called "I Rescued a Human Today."* It begins:

> *Her eyes met mine as she walked down the corridor peering apprehensively into the kennels.*

---

* Copyright Janine Allen, 2020 Rescue Me Dog; www.all-creatures.org.

I felt her need instantly and knew I had to help her.
I wagged my tail, not too exuberantly, so she wouldn't be afraid.
As she stopped at my kennel I blocked her view from a little accident I had in the back of my cage . . .

And it ends, after much care and consolation from the dog:

I was so fortunate that she came down my corridor.
So many more are out there who haven't walked the corridors.
So many more to be saved.
At least I could save one.
I rescued a human today.

The poem is an interesting choice for a dog, not least because it posits a world where the dog is in charge. Dog owners know that feeling of mild slavery rather well: the human–dog bond is stronger than ever, but our control on the leash not quite as taut as it was, the dogs rescuing us.

11.

The Art of the Floofiest

In the autumn of 2018 an exhibition called *Good Grief, Charlie Brown!* opened at Somerset House in the heart of London. Visiting was a marvelous experience, not least because it contained the clearest explanation yet of how a beagle who wasn't real became the most popular dog in the world.

The show launched at the perfect time. At *Peanuts*' peak fifty years ago, Snoopy became a global repository of love. He was a dog for all seasons and purposes, a dog both wise and cynical, practical and fantastical. His appeal is enduring and unmatched, and perhaps only today, through social media, has he met his true medium. Cartoons have always enabled us to say the things we find too complicated or unsuitable to say in normal human discourse. It was Snoopy's trick to express these things as a human writing as a dog thinking like a human.

Snoopy, or at least a version of him, began life as a small and utterly passive creature sitting sweetly on a cushion in a cartoon called *L'il Folks* in the late 1940s, and made his first appearance in the third *Peanuts* strip on Wednesday, October 4, 1950, getting soaked as he passed under a newly watered window box. Snoopy was first called Snoopy a month later, and within two years he was walking on two legs and expressing himself in thought bubbles. His first thought, which came to him on May 27, 1952, set the tone for a whole life, and indeed the life of all dogs: after a patronizing act from Charlie Brown, Snoopy wonders, "Why do I have to suffer such indignities?" Why indeed, for he may already have had a premonition of becoming the all-knowing, fairly selfish, reassuringly flawed and most enduringly merchandised cartoon dog in history. (But he was always dependent on the small humans around him to make this happen.)

Some 355 million people followed Snoopy and his friends each morning at the height of their popularity in the 1960s and 1970s. His adventures were syndicated to more than 2,500 newspapers, and in 50 years he appeared in 17,897 strips. (If you had to count something, that wouldn't be a bad thing to count.) His creator, Charles Monroe "Sparky" Schulz, worked almost every day of his adult life (he died in 2000), shouldering what he felt to be a great responsibility. People who bought newspapers expected a pleasurable and meaningful beginning to their morning, Schulz believed. He hoped that the sight of a little dog bouncing along would do the trick, and he was right: "Just thinking about a friend makes you want to do a happy dance," he noted, "because a friend is someone who loves you in spite of your faults."

Today the popularity continues in only slightly diminished form, for those who loved the comic strip as children found that in later life they couldn't shake it from their collective memories. Nor would they want to: *Peanuts* was a part of childhood that evokes only pleasure and no embarrassment, which is why, in this particular quarter of Somerset House, the grandest eighteenth-century building in London, almost all of my fellow visitors are adults, and almost all of them are charmed beyond caring.

It's not all Snoopy and Charlie Brown, of course. Lucy, Linus and the rest of the crew are scratched here too, all brilliant coconspirators in what *Vanity Fair* called "the longest-running meditation on loneliness, defeat, and alienation ever in popular American art." But it is Snoopy who leaps the highest with those happy-dance jump lines beneath him, and Snoopy who seems to understand more of life's mysteries than do all his fellow characters combined.

Charles Schulz describes Snoopy as having "a quality of innocence combined with maybe a little bit of egotism: you put those two qualities together and I think you have trouble." But you put those two things together and you also have a dog you can rely on in times of emotional need, and a dog you can hold up as a symbol of the age. Soldiers and astronauts took him to Vietnam and the moon, pinned and painted on their helmets and spacesuits. He was—and remains so, for he is no more dead than you or me—an emblem of fortitude, resentment, despair, hunger, greed, hope and deep love. He wants to do the best he can at a time of confusion and anxiety. He embodies an observation that Charles Schulz made of friendship in general: "Friendship isn't about who you've known the longest. It's all about the friend who comes and stands by your side in bad times."

The children with whom Snoopy shares his days are unlucky in love and at sports, for life is disappointing, and the things they hope for (a home run! cards on Valentine's Day!) seldom materialize. Amid this tiny carnage, Snoopy teaches the values of the imagination. He is ever the World War Two flying ace who will rid the world of evil, and the African lion hunter thrashing his way through the jungle. Can it just be coincidence that the ship which carried Darwin to his theories of evolution was called HMS *Beagle*?

Apparently, yes: Snoopy only became a beagle because his creator thought "beagle" sounded funny. His name had been suggested by his mother many years before Charles became a cartoonist; they already had a dog, and she thought Snoopy would be a good name for his successor. That's one origin story, at least; another relates how Schulz wanted to call Snoopy "Sniffy," but that name was already taken in another cartoon. (The Schulz family dog was a mixed-breed called Spike, and he was a bit of a celebrity himself: when he was thirteen, Schulz submitted a drawing of him for the illustrated newspaper panel *Ripley's Believe It or Not!* and it was accepted. Spike's special power was self-harm: he became famous for eating "pins, tacks and razor blades.")

Snoopy's talents were multilayered. He would don sunglasses and take on the persona of Joe Cool, a Fonzie-style college kid who did a lot of preening and leaning against walls. But he was just as happy being, in no particular order, the World Famous Attorney; the World Famous Member of the Emergency Rescue Squad; Dr. Beagle and Mr. Hyde; the World Famous Census Taker; and Blackjack Snoopy, the World Famous Riverboat Gambler. And would it come as a surprise to learn that Snoopy could also write and type? When the pressures of the world became too much for him, he turned his hand to literature, an early inspiration for at least one dog who came after him. In 1971, Snoopy published the elaborately titled *Snoopy and "It Was a Dark and Stormy Night."* This was a milestone, not least because his novel shoehorns all genres into one. As a curator at Somerset House pointed out, the book had mystery, violence, family intrigue, pirates, class struggle, a coming-of-age saga, a change of seasons, pathos, medical drama, a unification of separate narrative threads, romance, a kidnapping, an escape and a happy ending. Despite the absence of zombies, the book sold well.

The only person who doesn't like Snoopy is the Red Baron, his wartime foe in the air. And maybe his desert-dwelling brother Spike, who might covet his urbanity. But for everyone else the beagle has proved himself time and again to be vulnerable, sensitive and authentic. How authentic? So authentic that he can never remember his owner's name. In one strip, Charlie Brown returns home after a stay in hospital. He is delighted to be back, and immediately calls for his companion. He tells Snoopy he missed him, and would often lie awake at night thinking about him. After a pause, Snoopy concludes, "Now I remember! He's that round-headed kid who always feeds me . . ."

It's a cunning ploy, asserting dogged independence. A dog who can't name their owner can never really be said to be owned by one. More to the point, Schulz and Snoopy both understood that he was owned by his readers, which meant that everyone owned and loved him equally, and it was a love made new each morning. If you had always yearned for a dog, Snoopy became your ideal. If you already had one, you might have wanted a Snoopier one. And if you were a dog, let me tell you, you wanted Snoopy in your gang.

I'M NOT sure quite so many dogs wanted to be like Fred Basset. Basset was Britain's answer to Snoopy, which meant the cartoon strips had more rain, boxier cars and the occasional use (by the dog) of the word "mate." Fred Basset is what they used to call "sassy," often looking at the reader as he mocks his owners. But he's also a moper, and he often chooses the fireplace over an outing, not least because his body mass index tends to the Falstaffian. Unlike Snoopy, he's not a fantasist, and he's not inspirational; Snoopy went to the moon and Vietnam, Basset went to the greengrocer. In other words, he is indeed a basset hound.

But the most remarkable thing about Fred Basset is that he refused to die when his owner did. He was like one of those crazy dogs you sometimes read about who meet the same train in Italy every day in the vain hope that someone who died a decade ago will miraculously reappear with salami. Alex Graham, the Scot who drew *Basset* from 1963 until his death thirty years later, stockpiled about eighteen months of cartoons to run after his demise, but Basset fans in the *Daily Mail* and *Mail on Sunday* weren't sated. They wanted—nay, *demanded*—more Basset! Graham's daughter Arran Keith and the artist Michael Martin took up the challenge, gently updating the dog to keep pace with the modern world. In the recent strips, for example, children take pictures of Basset on their phones; when his owners eat out, they order calamari "in a crispy panko crust with sweet chili mayonnaise." The equivalent in *Peanuts* would be Snoopy atop his kennel listening to Billie Eilish.

Reassuringly, as the world accelerates around him, Fred Basset's attitude to life remains constant. Now in his mid-fifties, he is still "almost human," while never actually becoming less dog. He still has thought bubbles rather than speech bubbles, and he still thinks a hot date is a night by the fire. Unlike Snoopy, who appears both eternal and immortal, Basset is an anomaly; he's not particularly funny, and sometimes not funny at all, but like Green Shield Stamps and the labor exchange, he speaks of a simpler age; his appeal is reassurance and comfort, and sympathetic eyes. He's a true pet.

And he is narratively consistent. Snowy, from *The Adventures of Tintin*, is not consistent. For the first eight books, Snowy (or Milou, as he's called in the original French) appears in Hergé's graphic novels as a talking dog. Snowy's is a subtle and sometimes cynical role, modulating Tintin's behavior, chatting to other characters, being generally indispensable. Sometimes it's just engaging an internal monologue, an aside to the reader. But sometimes Snowy sets up an entire scene, as in the opening of *The Blue Lotus*: "How can a dog get a wink of sleep?" he complains with a pained face as Tintin's primitive telecoms machine crackles nearby. "Not a minute's peace since he fell for short-wave radio!" But then, after eight books, the cat gets his tongue and Snowy's role diminishes. Henceforth he will speak only occasionally, and then only to Tintin. Most blame his new reticence on the introduction in the ninth book, *The Crab with the Golden Claws*, of hard-drinking Captain Haddock, who takes on Snowy's wisecracking attributes; Tintin only needs one foil. What Snowy thought about this decision we never learn, because, like so many dogs, he remains a patient, loyal and long-suffering companion. But I bet he wasn't happy at all.

In printed cartoons, dogs *simply know more*, and they reflect the wisdom of dogs in the most compact way. In just a few inches, with directness and simplicity, they package truths about the human–dog relationship in a way that might otherwise take an entire essay. Primarily we are laughing with the dog, and principally we are laughing at the human. As newspapers diminish, it's just another truth we stand to lose.

ONE SEARCHES in vain for the all-knowing-dog cartoons of today. Snoopy and Snowy and Fred Basset are as cute as ever, but they don't raise the laughs like they used to. In their place we have something more bizarre but equally pointed, and with a far wider and more immediate reach: real dogs that increasingly just *look* like cartoons.

You will find them in the portfolios of famous photographers, and increasingly in the ultra-indulgent, excessively silly, and heartwarmingly life-enriching dog worlds that inhabit social media.

Emerging from the Charlie Brown show into the magnificent Somerset House courtyard, I rubbed my eyes in disbelief. Here, in a pre-planned gathering opposite the café, were cartoon dogs made flesh. Italian greyhounds and toy poodles, ten in all, but all of them wearing clothes. They were stars of their own framed tableaux, celebrities all over the world.

"The main person isn't even here yet," said a man with an elegant Italian greyhound in a pink dinosaur outfit—that is, a dog in a pink felt bodysuit with maroon spines down its back. The main person turned out to be a woman called Tess, and she arrived shortly afterward with two other Italian greyhounds called Winston and Twiggy. The dogs were dressed in coats like feather boas, one pink and one blue. Tess, who was dressed in a black leather jacket and ripped black jeans, was carrying a black tote bag with a question on it: "Darling Where Did All Your Money Go?" The next line reads: "My Dog's Wearing It."

There was a little gasp when Tess arrived, for her dogs are celebrities, and they have found a new, creative way of plotting a dog's journey through the world. Winston and Twiggy are stars on Instagram, where they have their own account and appear in all manner of boutique outfits and cute scenarios. Tess is a talented photographer, so her photos tend to be well composed with an alluring depth of field, and they may remind the viewer of classic shots of Hollywood starlets. Winston and Twiggy have a large and colorful collection of coats and hats, and spend a lot of time on beds and sofas, and they love the snow on Hampstead Heath. In most of the photographs they are looking directly at the camera, and neither of them look at all bothered by the attention or attire—indeed, they look born to it, or at least used to it. Twiggy looks like a Twiggy (if one takes the all-elbows, waif-like sixties model as inspiration) but Winston definitely doesn't look like a Winston (unless his inspiration is George Orwell's undernourished Winston Smith).

As is usual on Instagram, Tess has tagged the photographs of Winston and Twiggy (collectively @theiggyfamily) with many hashtags that group them with like-minded others in the hope of attracting more followers and greater approval. So we have #twiggythetroublemaker and #winnietheboss, but also #dogsinclothes, #adorabledogs and #longdoggo.

At the time of writing, #longdoggo had 41,414 posts, many of the photographs featuring whippets, lurchers, Russian wolfhounds and Afghans. But no breed is more anthropomorphized in this gang than the Italian greyhound. Perhaps it's their doleful eyes; perhaps, given their size, it is their pliability. Beyond Winston and Twiggy, one click on #italiangreyhoundsofinstagram or #iggylove brings up 370,000 posts of other Italian greyhounds also dressed in outfits and looking special. There are so many that they swiftly begin to look normal; we would do well to remember that dogs were never meant to wear clothes (or if they *were* meant to wear clothes, is it realistic to assume they would predominantly mimic human attire and dinosaurs?).

We just can't help ourselves, it seems, although this will always be an inadequate excuse.*

A search for #dogsofinstagram produces more than 145 million photos and videos, a great many of them standard pictures of charming puppies and dogs enjoying life to the full. But there are also many dressed in ridiculous outfits, and many doing tricks. There are certainly too many videos of dogs in glass cubicles in puppy shops, many of whom would have come from puppy farms. They are small and helpless, and already display disturbing behavior, such as scampering on the glass walls. Many of the most viewed videos purport to show dogs that are smiling, but the smiles look desperate. The comments below say such things as "That poor dog doesn't look that happy," and "This is so sad . . . take this down."

Even though @theiggyfamily was recently boosted by the addition of a third member, #cindythemilkdrop, and has more than 20,000 followers, the family dwells in a quiet part of town compared to the Pomeranian @Jiffpom (9 million followers), the chiweenie @tunameltsmyheart

---

* Despite its popularity on Instagram, the miniature Italian greyhound is not part of the recent designer dog trend—it just looks as though it should be. Though small and delicate enough to fit on a lap during a long carriage ride through Victorian England, when they were indeed favored by Queen Victoria, we first encounter the Italian greyhound in the ruins of Pompeii. They enjoyed a renaissance during the Renaissance, appearing in paintings by Giotto, Piero della Francesca and Pisanello, and securing their position as a status dog at royal palaces and the homes of noblemen. The American Kennel Club maintains that while the Italian greyhound retains its instinct for hare coursing, it also makes a splendid modern-day "decorative couch dog."

("the Underdog with the Overbite," a tiny cross between a Chihuahua and a dachshund with 2 million followers), and @marniethedog, a senior shih tzu who always has her tongue lolling out as if she's drunk (1.9 million).

The most popular accounts on Instagram do not treat dogs like dogs, but a bright and photogenic form of a rather pathetic baby. They too are Snoopy-like objects of love, but they are more fantastical than cartoons: often they are animals made to look like anything they're not. Jiffpom has one of those faces which makes her look as if she's smiling. She has a vast wardrobe, and sometimes wears glasses to look studious, and sometimes wings, and often hats and shoes. Jiffpom also appears in a large array of human situations (eating at a restaurant, sitting at a workbench), which means Jiffpom is often required to be on her hind legs and to pose with celebrities, almost all of whom mimic Jiffpom by putting their tongues out. Needless to say, Jiffpom has her own merchandise range, including jigsaws and soft toys; in the photos, it is often difficult to tell which are the soft toys and which is Jiffpom.

Then there is @tunameltsmyheart, who, on account of his overbite, looks consistently worried and gormless. You can buy a photo calendar with @tunameltsmyheart's most arresting looks, but it would be a bit like buying a photo calendar of a person with an unusual facial disability and then laughing at that. Here he is as Father Christmas; here he is wearing a fleecy red wrap with a large bird, possibly a toucan, sewn onto the back, so that it looks as though he's giving the bird a piggyback; here he is in several wigs, which make him look like a mermaid and a pharaoh. A freak show, in other words, not in dank Victorian London, but hugely popular on all devices globally.

If I have uncovered one persistent theme in these pages, it is surely this: nothing in our buoyant relationship with dogs is entirely new. Emotional attachment doesn't change very much from century to century. The affectionate portrayal of our dogs is consistent from rock paintings to Landseer to Charles Schulz, and it is certainly consistent on Instagram; we may not always have welcomed the dog so warmly into our living rooms, but we have always depicted them visually as rewarding, comforting, useful and amusing. And the more we may portray them in our own image, or evidently in the service of our goals, the happier we seem to be.

THE DOGFATHER of the Instagram look is William Wegman. You must have seen his work—his beautiful silvery gray-brown Weimaraners got up in all manner of human garb and in all types of human scenarios. Here they are in a wig or an overcoat or a hat, and now as a hockey star, or George Washington, or a character from *Star Trek.* They sit on chairs, on cubes, on Grecian pedestals. They have lampshades on their heads and hoops on their noses. It's not a cruel act, just a preposterous one, and it's the unnaturalness that encourages curmudgeonly admiration. But a curmudgeon you should remain, for these are *dogs.* Evidently pliant and never seemingly distressed, they are nonetheless performing tricks not for their betterment but for the betterment of their demanding master. There is money to be made, and they are only one cultured step away from the freak show and the circus act with poodles.

Wegman began working with his first Weimaraner muse Man Ray in California in 1970. He noticed that the dog would whine when he wasn't in front of the camera. "He wanted in and he made that clear to me," Wegman explains in *Being Human*, the book of his greatest hits. "Because he could be so calm and focused it was possible for me to work with him like any other object or prop. I could build a picture around him." Wegman says he was initially wary of making Man Ray the subject of "cuteness, beauty and anthropomorphism," but after a few years, when he started working with a large-format Polaroid camera, things changed. "With two Polaroids in particular—the unabashedly beautiful *Double Portrait* [Man Ray on a sofa peering out from beneath a printed photo of himself] and the dangerously anthropomorphic *Man Ray and Mrs. Lubner in Bed* [as it sounds—two dogs looking postcoital with their heads on pillows in a log cabin] . . . my resolve was shaken and ultimately abandoned." When Man Ray died, Fay Ray took his place in the studio; and when she had eight puppies, a photogenic dynasty was born.

The highly regarded photography critic and museum curator William A. Ewing rigorously takes Wegman's corner on this. In fact he takes the dogs' corner, playfully suggesting that most of the ideas for poses originate with them. They find humans wildly amusing and never tire of parodying their supposed masters; Wegman merely becomes the "faithful photographer." "Still, I would not like to leave the impression that William Wegman's role is merely to snap the shutter," Ewing explains. "He is listened to respectfully, and his ideas are often adopted . . . It is best to think of the production as highly collaborative.") That said, we are obviously yet to hear from the dogs themselves, and they may have quite a different story: they look settled enough in the photographs, and I am sure they are well cared for. But looking at the pictures always makes me feel a little queasy; I'm not sure all that posing would have been the dogs' first choice of career.

If I were a dog, I think the person I'd really like to photograph me is Elliott Erwitt. You'll have seen his work too, even if his name is unfamiliar. A Chihuahua in a knitted woolen hat stares inquisitively next to the legs of a woman in boots and the legs of a Great Dane; a boxer poos on a wide city pavement with a "you and whose army?" look on his face; a dog jumps for joy in Paris next to his flat-footed, raincoated walker.

To get a dog to jump you have to bark at it, Erwitt says. "You have to speak their language." No one speaks it more fluently. His photos show dogs as they live, with every emotion from ecstasy to despair. No dog wouldn't recognize him-or herself at his shows or in his books, and the most common reaction will be delight. There are some really scruffy and stinky mutts in with the aristocrats, and not all of them are friendly faces. But they're all equal and special in front of his lens, and Erwitt has succeeded in investing his many dogs with a universal attribute: dignity.

"What superb photographs his are," wrote P. G. Wodehouse in a collaboration with Erwitt published in 1974 called *Son of Bitch*. "It does one good to look at them. There is not a sitter in his gallery who does not melt the heart." In the same article, the novelist expressed his thoughts about dogs and comedy, declaring that some dogs simply have a sense of humor and some dogs simply don't. "Dachshunds have, but not St Bernards and Great Danes. It would seem that a dog has to be small to be fond of a joke. You never find an Irish wolfhound trying to be a stand-up comic."

I once met Erwitt at a Leica event in Germany and asked him, "Why so many dogs?" by which I meant "why so many dogs to the exclusion of other animals such as cats or horses or cows?" Erwitt replied, with a smile that may passably be described as enigmatic but more accurately was just annoying, "Why so many humans?"

In the introduction to his fat book on dogs entitled *DogDogs*, the photographer reasoned, "I don't know of any other animals closer to us in qualities of heart, sentiment and loyalty. Some people say elephants come close. Personally I find elephants too bulky . . ."* There may be another reason: an unphotogenic dog has yet to be born. The faces of dogs do not age in the manner of humans: they do not wrinkle, they do not look excessively forlorn, and they are not tempted by surgical intervention; the worst that happens is a little graying around the jowls.

Erwitt's interest in hounds began as a teenager in the

* Robert Adams, the great photographer known primarily for his landscape work, had another interpretation of a dog's appeal. "Artists live by curiosity and enthusiasm, qualities readily evident as inspiration in dogs," he writes in *Why People Photograph* (1994). "A photographer down on his or her knees picturing a dog has found pleasure enough to make many things possible."

mid-1940s, when he adopted a mutt with life-threatening canine distemper. The dog was named Terry, and as his appearance grew increasingly disheveled, "his intelligence and sensitivity grew." He didn't live with Erwitt full-time, but maintained his dignity on the Los Angeles streets, accompanying the postman on his rounds, chasing a bicycle or two, occasionally dodging the traffic to visit the photographer's mother several miles away in east Hollywood.

Erwitt later adopted another dog, a cairn terrier named Sammy. Sammy was a working dog, insofar as he worked for Elliott Erwitt. This wasn't quite in the full-time manner of William Wegman's dogs, who clearly should have got at least half the photographer's income; Sammy was to be employed merely on one project, albeit an ambitious one. The plan was for him to wee at the foot of every major structure in the world, from the Arc de Triomphe to the Pyramids of Giza, from the Lenin Mausoleum in Red Square to the Forbidden City in Beijing (in this case *very* Forbidden). The project was ongoing, but awaiting final funding, when Sammy passed away.

These days, through the midwifery of technology, we are all Erwitts and Wegmans, albeit with less artistic ability. The smartphone and the cloud make us all responsible for our own ethical legacies. Millions of "likes" each day cannot but make an impression on our minds of how a dog should look, and they make the village dog look distinctly plain. You have a regular Labrador or Staffie or ordinary mutt? Where, unless you dress them up, are the clickable endorphins in that? Perhaps what matters most about these images is their ubiquity and their immediacy, and the new use a human makes of dogs to boost their own attractiveness or perceived likability, digital or real. Like so much social media, looking at it does not leave you feeling very good about yourself or particularly hopeful for the fate of the world.

The canine behaviorist Alexandra Horowitz is worried, above all, about the possibility of losing the very thing that makes dogs dogs, and about the risk of judging them against a human ideal. "The omnipresence of my favorite subject has begun making me grumpy, not elated," she observed in the *New York Times* in 2018. "As dogs themselves produce a profound anti-grumpiness in me, I began to wonder why. Why can't I stand to look at one more photo of a 'funny dog'?"

The answer was that social media and the entertainment industries have made dogs into "furry emoji," a shorthand replacement for emotion and sentiment. The act robs the dog of dignity, and, observes Horowitz, "each representation diminishes this complex, impressive creature to an object of our most banal imagination . . . it is degrading to the species."

Of course she's right. Whether it is *damaging* to the species is another matter, for humans have been dressing up dogs in bonnets to please the eye since the Victorians. In 2015, Horowitz combined forces with Julie Hecht, a psychologist at City University of New York, to examine the factors that, these days, make humans fall for one dog rather than another. Some of this (as we have seen with the labradoodle and its friends) comes down to fashion. But were there other elements that might constitute a new theory of visual natural selection, elements that will affect a prospective new dog owner at a rescue shelter?

At Barnard College in New York City, 124 students were asked to examine eighty pairs of photographs of mixed-breed dogs on a computer screen and indicate which ones they liked best. Each pair featured two nearly identical pictures of the same dog, but contained one physical characteristic that had been deliberately manipulated. The researchers had identified fourteen different physical attributes to be investigated, each relating to an established psychological theory of attraction. Would the size of a dog make a difference, perhaps? Or the size of a dog's eyes? Looking at a sample of the photographs in the published study it is very difficult to tell them apart. So we choose the ones we like best on instinct and neural programming.

The results were fascinating, if perhaps predictable. There was a marked preference for features associated with infants, including large eyes and eyes set widely apart. The students also preferred certain attributes that made the dogs appear more human, including colored irises and an expression that suggested a smile. There was no preference for physical size or symmetry. The researchers had shown what regulars on Instagram already knew instinctively—the more a dog looked like a human, particularly an infant one, the more we liked that dog. It was the science of cute.

The study, and our newly trained eye for what brings most approval on social media, raises a handful of interesting questions. If the factors motivating the selection of dogs relies heavily on aesthetics, are we in danger of forgetting what it was that made a dog such an indispensable and companionable animal in the first place? Is their dogness being replaced by our humanness? And does this matter?

BUT THEN again, lighten up! Surprisingly, the best antidote to Instagram may well be found on Twitter, specifically an account called WeRateDogs (@dog_rates). Here all the dogs are stars, and ordinary dogs without clothes tend to be the biggest stars of all (an occasional bandanna, bow tie or pair of reading glasses is allowed). The notion is simple: a photo or video of a dog is submitted for careful consideration, and almost inevitably the dog emerges with kudos. The dog is supposedly rated out of ten, but always achieves more than ten, sometimes twelve, thirteen or even fourteen out of ten. The principal judging criterion appears to be that they are a dog. WeRateDogs adds many dogs to its feed every week. One week in 2019 they added:

> This is George. He just graduated with a double major in snuggability and good sits. Wonderful job George. 14/10, please congratulate him

> This is Charlie and Maverick. Charlie had his eyes removed due to glaucoma, but then Maverick came along as his little helper. Now they're doing amazing. Both 14/10

> This is Lucy. She hasn't made it out of bed. The world just seems like a lot today. But she knows how much her human needs her. 13/10 you got this Lucy*

You lose a little without the photos, but you may still be reminded of the summation by John Caius, some five hundred years earlier, of a really good spaniel: "It is a kinde of dogge accepted among gentles . . . they will not only lull them in their lappes, but kysse them with their lippes, and make them theyr prettie playfellows."

But one would have to have a hard heart indeed to suggest that most of the doggos on the site are *not* worth twelve, thirteen or even fourteen out of ten, particularly the large number featured each month who need financial support for vet fees. People see their plight and send in money or offers of support, and those who have been helped often reappear on WeRateDogs a few months later looking healthier, accompanied by their grateful owners, who are keen to stress that their photo was genuine and their appeal not a scam.

WeRateDogs has more than 8 million followers; perhaps you are one of them. It began in November 2015 as a comedy outlet for Matt Nelson, an eighteen-year-old student at Campbell University in North Carolina. He knew that dogs were always a huge hit on social media, and he says he wanted to "utilize dogs" to help his comedic slant on life to reach more people. His slant was not snide or

* George tweet: March 7, 2019. Charlie and Maverick tweet: March 18, 2019. Lucy tweet: February 21, 2019.

snarky like a lot of other Twitter comedy accounts, but absurdly positive: dogs had rarely been so uncritically love-bombed. And for reasons that even its originator can't explain, things took off extremely fast (the only logical explanation is that people love dogs and anything that makes them even more lovable will always go down well). A dog joke about one dog tends not to be dog-specific; if you're a dog person you won't take much convincing that it's also really *your* dog in the photo. Besides, since the passing of Grumpy Cat, the fans of WeRateDogs believe that its dogs are one of the few things that give social media a real and rare purpose.

WeRateDogs certainly confirms the nonsense that dog lovers direct at their pets. Alexandra Horowitz has compiled a list of overheard "conversations" New Yorkers have with their dogs on the streets—a mix of sweet nothings, questions and instructions—and they are often surreal. "You're so cute and so smart," a woman said to her goldendoodle. "And worth money! I could marry you." A reluctant dog was encouraged by her owner in the way she might otherwise talk to her kids: "You can sit all you want when we're home." And, most convoluted of all (and perhaps a classic case of what a psychoanalyst would call transference), is an owner addressing a shy dog avoiding the friendly advances of another dog: "You're embarrassing yourself!"

I don't think these are anything new: I think we have always conducted these monologues, albeit not always to goldendoodles. What has changed is our unashamed willingness to talk to our dogs this way in public and out loud.

Social media has emphasized what the interconnected digital world has done from the beginning: shown we are not alone with our madnesses.

It was only a matter of time until Matt Nelson's affectionate/tortuous wordplay—"Doggos," "boopability" and "floof"—would soon spawn WeRateDogs mugs and T-shirts, and a WeRateDogs calendar and board game. The object of the game is simple enough: to prove that your dog is the goodest of them all. Or as it says on the promotional material: "After a busy day at the shelter finding some new best friends, you've come home with the floofiest, sassiest, speediest, most boopable doggos imaginable—and then you hear that a dog show is being hosted right down the street!" One of the game cards features a dog called Thor, who is rated thirteen for "sass," but an extraordinary twenty-two for "goodness." You must assume that Thor would hammer every other dog into submission; happily the game isn't that simple.

A little cloying for you, perhaps? A little annoying? Wish you'd thought of this lucrative idea yourself? After almost two years of running the account as a sideline, Nelson quit university in 2017 to rate dogs professionally. In other words, he found a way of monetizing his doggos. His T-shirts and baseball caps say things like "Tell Your Dog I Said Hi" and "Oh H*ck" (that's just like "Heck," but more modest and floofier). These are not particularly deep or clever things, but the warmth is endearing and the merchandise reroutes a digital community into a real one. "Nobody's going to stop liking dogs anytime soon," Nelson told *Money* magazine in 2018, "so there's some job security there." WeRateDogs is still headquartered at Nelson's parents' house near Charleston, West Virginia; his father, a legal professional, advises on matters of finance.

Despite its success, or perhaps because of it, WeRateDogs has—as astonishing as this may appear—received criticism for not being scientific enough. Some people believe that the dogs are overpraised. The most famous of the naysayers is a man called Brant Walker (@brant), and in September 2016 his distrust of the rating process made him virally famous.

The story began as a mild troll: "Your system sucks," Brant tweeted. "Just change your name to 'CuteDogs.'" Matt Nelson promptly replied, deliberately misspelling his accuser's name: "Why are you so mad Bront."

Brant Walker kept it coming: "Well you give every dog 11s and 12s. It doesn't even make any sense." And then Matt Nelson responded with the phrase, "They're good dogs Brent."

For a while, "They're good dogs Brent" became one of social media's most overused (almost) meaningless phrases, quoted and adapted in myriad ways, and reproduced on badges and T-shirts. You tweeted your objection to a new governmental transport plan, say, and within minutes someone would tweet back, "They're good road policies Brent." The *Washington Post* named "They're good dogs Brent" as "the best meme of 2016," and quoted Matt Nelson's claim that the screenshot of the original Twitter conversation has been reproduced 35 million times. Matt Nelson and Brant Walker are both happy to have been involved in this public spat, grateful for the extra followers. When Walker got married, the wedding cake was decorated with Marge and

Homer Simpson, and beside them stood the Simpsons' family dog Santa's Little Helper, who was supporting a placard that read, of course, "They're good dogs Brent!"

In the summer of 2018, Brant Walker's household acquired an elegant Havanese dog called Charlie. It seemed only natural to send it to WeRateDogs for a verdict. The verdict was "This is Charlie, dog of Brant. He naps in a pineapple and really likes French Fries. 14/10 you've got a good dog, Brent."

What would Sir Edwin Landseer and Charles Schulz make of this fuzzy nonsense, given the chance? My dog lover's instinct—that part of a human heart that's forever melting in the presence of canine absurdity and cuteness—suggests to me that they'd be following Matt Nelson on Twitter, and they'd be buying the merchandise. The technology changes but the sentiments hardly do, for what is a stupid post on Instagram if not an update of a lordly depiction in oil? Our love of dogs, and the desire to portray them at their best—their most daring, their most curious and beautiful, their most inquisitive—remains a constant in our relationship. One immerses oneself in these images, and we find it difficult to imagine a time when we didn't try to capture them in that frozen moment when they behaved just the way we wanted them to, or a time when they weren't the loving focus of our unconditional and unfettered devotion.

12.

Inglorious

In June 1939, the popular monthly British dog magazine the *Tail-Wagger* was so full of doggy optimism that newsagents probably had to restrain it from bouncing off the racks and licking you in the face. There was good news in the very first article: the "dog tax" was not being raised by the chancellor of the exchequer as owners had feared; the government already received more than £1 million from taxing dogs, and that, in the eyes of the *Tail-Wagger*'s influential editor, was quite enough.*

* The dog tax, in the form of the dog license required for all dogs in England, Wales and Scotland, was abolished in 1987, when it stood at 37 pence per dog (but the stipulation still exists in Australia and some parts of the United States, and also in Northern Ireland, where the cost is £12.50 annually). The idea of the license was first introduced in 1871 as a protection against stray dogs and the threat of rabies, and was amended in 1906. By the time of its abolition it was reported that only about half of all owners paid the annual fee. Universal mandatory microchipping has generally

The next spot of joy came with news about Peter the Irish terrier from Horn Lane, Acton. Peter would trot through rush hour traffic, cross busy junctions, pause at a Belisha beacon and arrive at Acton Station to pick up a copy of the *Evening News* for his master, Mr. F. A. Gibbon. "At one time he [also] used to fetch the racing and lunch edition of the *Evening News*," the magazine reports, "but of late these have been delivered by car for greater speed." Elsewhere there was a report of training advances for guide dogs for the blind, and the results of the annual *Tail-Wagger* scholarships (the fact that the Casebrook Scholarship to the value of £12 and 12 shillings had been awarded that year to Mr. A. R. Casebrook didn't seem to raise eyebrows). Among the adverts, Cooper Dog Remedies explained that unless you buy their product then "Dog can't win when it's dog v dog-flea," while Ambrol Milk Food was perfect for "puppies, bitches-in-whelp, sick dogs, and poor feeders" and in fact all other dogs too. On every page there were hounds looking happy: the future could not have been brighter. But look at that date again.

The years to come were to prove as traumatic for dogs as they were for humans. In fact, the *weeks* to come would bring a tragedy to dogs on a scale never experienced before or since.

In the first four days of the war, an estimated 400,000 domestic dogs and cats were killed voluntarily in London, an act dubbed by the historian Angus Calder as "a holocaust

---

replaced the license as a form of registration.

of pets." (He also describes how slain pets "lay in heaps" outside vets' surgeries.)

A holocaust of pets? Four hundred thousand? How can this possibly be? And how can this miserable breakdown of the indestructible bond between humans and their companions be afforded so little space in today's collective memory? Perhaps one's horror at the number provides an answer to the question: the thought of the massacre is almost too much to compute, and certainly too much to bear.

The number, which represented about a quarter of all cats and dogs in Greater London at the outbreak of the war, was confirmed by the RSPCA and the official wartime report of the Royal Army Veterinary Corps; in fact, the RSPCA president, Sir Robert Gower, suggested the total could be almost double this, at 750,000. The figure is all the more astounding because relatively little documentary evidence exists to add much detail to the figures. One example, however, from the art critic Brian Sewell, provides a chilling account of how routine the death of a pet could become. "Robert shot him and left his body on the beach for the tide to sweep away," Sewell wrote in *Outsider*, the first volume of his memoirs. Robert was his stepfather, the beach was Whitstable, the dog was Prince, his Labrador. "Packed among the suitcases in the car, I saw Prince led towards the sea and heard the shot. I did not cry, as I would now . . ." Another example appeared as a memorial notice in the *Tail-Wagger*: "Happy memories of Iola . . . sweet faithful friend, given sleep September 4th 1939, to be saved suffering during the war. A short but happy life—2 years, 12 weeks. Forgive us little pal, you were too nervous to be sent away. Au Revoir. Terribly missed by all at 6, St Ives Road, Birkenhead." And so this was us: the softhearted, British, dog-loving nation forced to destroy the very things we adored the most, horribly and literally killing our friends with love.

What led to this calamity? It seems that at least one cause of the slaughter was officially sanctioned. Just before the war began, the Home Office Air Raid Precautions Department drew up a special booklet for animal care, advising the removal of animals to the countryside. Failing that, it advised those facing enlistment or a similar dislocation, "If you cannot place them in the care of neighbours, it really is kindest to have them destroyed." The booklet contained a full-page advertisement from Accles and Shelvoke Ltd, a company from Birmingham selling its Cash Captive Bolt Pistol, and a photograph of the gun was accompanied by simple text: "Provides the speediest, most efficient and reliable means of destroying any animal, including horses, cats and all sizes of dogs."

Dogs were forbidden from public air-raid shelters, and at least one enterprising firm advertised its gas-proof kennels in dog magazines. Animal welfare organizations built their own special refuges, including one in Kensington Gardens sponsored by the National Canine Defence League. At times there appeared to be as many dog charities as there were breeds (in addition to those already mentioned, Our Dumb Friends' League and the People's Dispensary for Sick Animals of the Poor were particularly prominent), and, before they were overwhelmed by a combination of demand, desperation and internal disunity over how best to proceed in the crisis, they all attempted to save as many dogs as they could. Even as the RSPCA was destroying hundreds of animals a day, it campaigned for a moratorium: "This country has not reached a stage when the wholesale destruction of household pets is necessary," it claimed in its booklet *Feeding Dogs and Cats in Wartime.*

But it didn't seem to help. Companies such as Harrison, Barber & Co, "slaughterers and fat renderers," were already promoting their services to the Home Office for the beneficial by-products of animal disposal, which included fur, soap, fats, glue and fertilizer. (There was also distressing information from the dog welfare charities themselves on how efficiently they could dispose of pets; it was terrible destruction on an industrial scale, willful if not entirely heartless. Battersea Dogs Home used "electrothanaters" to kill one hundred dogs an hour; clinics managed by the Canine Defence League managed fifty dogs an hour by electrocution, chloroform or the injection of acid.) The fairy tales from the Metro-land brochures of the early 1930s—the confident family in the photos with the fluffy white terrier on the greenest of lawns—had turned to the cruelest of urban nightmares.

Some of the euthanasia was instinctual. A shared memory from the last war—starving dogs roaming the streets—was sufficient to suggest that a humane execution was preferable to witnessing a drawn-out period of emaciation (we don't have these village dogs in urban Britain, some reasoned; this isn't rural Greece). The Oxford historian Hilda Kean observes that the cull also reflected the changed role of pets in the home: although generally less pampered than they are today, dogs were also considered less "useful": as their role in security and hunting diminished, the more they became a luxury (and therefore expendable); the same applied to the cat's role as a defense against vermin. With food supplies dwindling, owners considered their new responsibilities: for humans, most domestic dogs were simply not pulling their weight. The RSPCA did its best to allay an owner's guilt: "Potatoes are plentiful, and if you put in extra tubers when digging for victory you will not have it on your conscience that . . . space is being taken for food for your animals." But the panic continued. In any wartime chain, dogs were increasingly lowly souls.

In September 1939, the comforting impact of dogs on the mental well-being of their owners was barely considered, alas, and nor was the effect on national morale. There was no "phoney war" for pets; mass and willful death was leaving blood on village doorsteps long before the Allies flew in anger. No one was a hero, neither pet nor person; no statues celebrate their valor. But this, alas, was indeed how we once treated our dogs in the perceived effort to save ourselves. It was the biggest organized act of cruelty to dogs within memory. Today I think even the hint of pet destruction at a time of national crisis would be met by revolution.*

---

* There were heroes, of course: a great many dogs were involved in the war effort as aids to detection, transport and what may best be described as comradeship, a like-minded sympathetic soul at a time of crisis and self-doubt. And the dog population did achieve at least a modest correction with

• • •

BUT EVEN in the darkest hours, dogs were dogs; they were still up to mischief, and they found their own way to boost morale. In other words, the Sieg-Heiling dog story is true. Jackie was half Dalmatian and half another dog, and in the only photograph that exists of him, which was taken on a balcony in the sun, he is actually wearing sunglasses. He is a maverick with a hint of mystery, but the mystery only unraveled seventy years after the photo was taken.

The German historian Klaus Hillenbrand first came across the Sieg-Heiling dog in 2010. A tip-off led him to microfilm at the German Foreign Office, where he found the story to be even more bizarre than he was expecting. "There are very few things you can laugh about, because what they did was so monstrous," Hillenbrand reasoned about the Nazis. "But there were two or three dozen people discussing the affair of the dog rather than preparing for the invasion of the Soviet Union. They were crazy."

The trail starts in January 1941, when the German vice-consul in Finland, a man named Willy Erkelenz, heard rumors that a dog in Helsinki would respond to the name Hitler by raising his paw in a salute. Finland was

---

the mass evacuation from Dunkirk in May and June 1940. French dogs were regarded as refugees fleeing persecution, and a fair number made it across the Channel to safety. Arriving in England, however, many of these would only respond to commands in French.

friendly to the Nazis at this point, and Jackie's salute would have remained an amusing party trick had the Third Reich not taken it as a grave insult.* As it was, restricting the trajectory of Jackie's paw and prosecuting Jackie's owner became a top priority.

The dog's owner was Tor Borg, a manager at a pharmaceuticals distribution business. When summoned to the German embassy to explain the rumors, he claimed that only his wife called the dog Hitler, and that the Sieg-Heiling began and ended not long after Hitler came to power in 1933; he said that Jackie would never dream of raising his paw in mockery ever again. The Germans didn't believe Tor Borg, but after three months of sleuthing they failed to find a single witness who would testify against him. Threats to curtail the supply of drugs to Borg's company came and went.

Tor Borg died in 1959, but we have no idea what happened to Jackie. There is no bronze memorial to the talented dog (and if there ever were, should Jackie be paw up or paw down?). We still can't be certain whether the dog was male or female. According to Klaus Hillenbrand, there is at least one important lesson to be learned from this episode: "The dog affair tells us the Nazis were not

* Humor not being one of the Nazis' strong suits. It's unclear how much Hitler himself knew of Jackie's exploits. He had famous dogs of his own, one of which—a German shepherd called Blondi—he killed by testing out a cyanide capsule before he killed himself in his bunker. Blondi's puppies were also then shot by Hitler's cronies, as were Eva Braun's two dogs and the dogs belonging to other aides. In effect, the slaughter of animals once regarded as pets—albeit pets burdened with wearisome layers of propaganda—was one last act of cowardice.

only criminals and mass murderers, they were silly as hell."

Should you require further proof of this, I would direct you to the exploits of Frau Schmitt, headmistress at Tiersprachschule Asra, a school established by the Nazis to teach speech to German mastiffs. And speak they apparently did, even if the words were all too predictable. When one of the dogs was asked, "Who is Adolf Hitler?" he answered (without thinking overly long), *"Mein Führer!!"* In 1943, Professor Max Müller, a supporter of the school, wrote an academic paper on the possibility of using the mastiffs in the war effort, not as fighting dogs, but in a more learned and intellectual leadership role.

This story comes courtesy of Dr. Jan Bondeson, a senior clinical lecturer at the School of Medicine at Cardiff University. The doctor also notes that the "sinister" Professor Müller was a keen supporter of several other investigations into the extracurricular lifestyles of German dogs, including the laborious work of a fat dachshund named Kurwenal, the dog who "barked the alphabet." Active in Weimar in the 1930s, Kurwenal became so famous that he had his own biographer, who was—fairly predictably, given the scenario—called Otto Wulf. The dachshund just barked all the time, and must have been such an absolute nightmare to live with that its owner, a certain Baroness von Freytag-Loringhoven, decided he should be put to some financial use. He would bark once for *A*, twice for *B* and so on, although he tended to double-back when he got to *M*, so that *Z* only merited one bark. Whatever. Kurwenal expressed no particular opinion on Hitler, but he stated a preference for Goethe over Schiller, and could identify a soliloquy from *Hamlet*. Once, when presented with a teddy bear as a gift and asked whether the bear looked appealing, he replied (by barking hundreds of times), *No, he looks horrible!*

(It's probably worth mentioning Kurwenal's famous Great War German predecessor Rolf, an "educated" Airedale terrier skilled at tapping out letters on a board with his paw. As well as being a militaristic patriot (he wanted to join the army to fight the French), he was also considered an intellectual. He was, as Dr. Bondeson notes, also a keen letter writer, and he "successfully dabbled in mathematics, ethics, religion and philosophy." I think the key word here is "dabbled."

The symbolic power of dogs is seldom more visible than at times of national vulnerability. Hitler chose to be accompanied in public by a purebred German shepherd, and when his catastrophic ambitions died, so necessarily did the dog. Churchill was often portrayed with a British bulldog, each as stout as the other, the dog often drawn sporting a Union Jack waistcoat; morale-boosting posters went further, grafting Churchill's head onto a bulldog's body, or a cigar-chomping bulldog's head onto Churchill's. During the war, the dog Churchill actually owned was a less impressive fighting force—a small brown poodle named Rufus. Rufus somehow failed to feature on the posters.*

---

* The Japanese scholar Aaron Herald Skabelund, a professor at Brigham Young University in Utah, notes that it is impossible to tell the story of a country's social and military development over the last two centuries without consideration of the changing role of dogs within society, suggesting

The hijacking of dogs for propaganda was nothing new, and not all the associations were positive. You didn't, for instance, want to be a dachshund during the First World War. In his memoir *A Sort of Life*, Graham Greene described a wartime day in which "a dachshund was stoned in the High Street" in Berkhamsted, Hertfordshire, where he grew up. Similar fates met the dogs in the streets of Manhattan and Chicago after the Americans entered the war in 1917. One hardly needed to imagine what sort of brutalism was being unleashed upon the world if such acts were being perpetrated on defenseless animals at home. Similarly, cartoons depicted a dachshund being shunned by other dogs in a park, and drooping and bloodied in the mouth of a bulldog. The propaganda had a crushing effect: in 1913, 217 dachshunds were registered in Britain, but in 1919 there were none. In the United States only twelve officially registered German sausage dogs survived the Great War, and there was a campaign by their owners to rebrand them "liberty dogs."

Hilda Kean has observed that as soon as the Second

---

that "imperialism shaped the world of dog breeding and dog keeping as we know it today." He also cites the fascinating "Dog Map of the World" that appeared in the *Illustrated London News* in 1933. Out of the seventy breeds recognized by the (English) Kennel Club and the American Kennel Club, thirty-eight came from the British Isles. Only three came from the United States, and a dozen or so from Germany, France and the Low Countries. According to the map, the Akita has not yet set its paws upon Japan, while in Mexico only the Chihuahua merits attention, and in the whole of Australia only the dingo. The imperial dog world of the nineteenth century was essentially European, and in the earliest dog shows those breeds not considered to have come from the British Isles were grouped in one category: "Foreign."

World War was declared in September 1939, the British press reveled in the comparable fates of two dogs to contrast the differences in British and German decency. Sir Nevile Henderson, the British ambassador to Germany, had returned to London with Hippy, his beloved Alpine dachsbracke. The fact that Hippy was not a dachshund, and had slightly longer legs than a dachshund and a fuller, wider body than a dachshund, and a countenance that was as much bloodhound as dachshund, did not deter the British press from calling him a dachshund. But this dachshund was not reviled, he was worshiped. Sir Nevile's love of Hippy was a sterling British example of how much respect we show to our pets in these grave times, even if the pets are German, or in this particular case Austrian (and this wasn't true anyway, given the mass slaying outlined previously). As it transpired, Hippy's fate was hardly lovely: he disappeared into quarantine as soon as the ambassador's plane landed, and when he was reunited with his owner six months later he was a much weaker and less resilient hound all round.

A direct comparison was made with the fate of a chow called Baerchen. Hapless Baerchen had the misfortune to be Joachim von Ribbentrop's dog when he was stationed as ambassador to Britain, and he remained at the London embassy when his owner returned to Germany as foreign minister in 1938. When the rest of the embassy staff flew back to Germany at the outbreak of war, no one apparently deemed it necessary to reserve a place in the hold for Baerchen. "That's what Britain is fighting," the *Daily Mirror* proclaimed, "the inherent brutality of Nazi-ism, that has no justice or human feeling—even for its pets." But irony abounded. Many appalled and kind-hearted *Daily Mirror* readers vied to adopt Baerchen, and when he found his new warm home he enjoyed a more pleasant life than Hippy.

We should remember these calamities. From the top of Parliament Hill we may still discern the smoke that once swirled around St. Paul's. Dogs today are generally so well cared for, and so well protected and pampered, that we would be wise to recognize their current lives for what they are: invaluable expressions of freedom.

Okay, sermon over. Finally, we are gathered here not to mourn dogs but to praise them. How best to remember your dog? What words can possibly sum up what an old dog means?

13.

Born a Dog, Died a Gentleman

On the northerly fringes of London's Hyde Park, as the road inside joins the Bayswater Road, there's a gatekeeper's lodge with an extraordinary garden. Within it lie around three hundred gravestones, each the size of a menu in an old-style Italian restaurant.

Here there are memorials to Turk, Little Nora, Darling Sammie, Dear Little Minnie and Sweet Little Leo. "To dear little Josie," one of them proclaims, "in loving gratitude for his sweet affection, until we meet again." Ten feet away: "In loving memory of Chum, my faithful and loving poodle," and "My own Bob, for 5 years the beloved and devoted companion of Mr F.M. Dican."

The Victorians knew how to do this thing well, these agonies of hindsight, but the intervening years have not been kind to the inscriptions or the stones, and the fading and crumbling make interpretation tricky. One can still make out:

Scamp, run over, 29th September 1894.
Here lies Tip, Sept 8, 1888.
In loving memory of dear Chin Chin, a perfect dog.
In memory of my darling little dog Pickles, who died on January 31st 1914, my faithful little friend and companion for 12 years.

We bury dogs in a human way, because it's the only way we know; in death, dog and owner have become one.

Alas, it is not possible to view this woeful corner of Hyde Park on a whim. To do so one must either be friends with the lodge keeper or sign up for a twelve-person private tour organized by the Royal Parks. If the latter, the cemetery visit is the finale of a broader ramble, beginning at the animal war memorial on Park Lane (a tribute to every animal, from pigeon to elephant, that helped in every war, from Crimea to Iraq), and taking in the site of the Tyburn hangings and the wonder of London's plane trees. Jonathan the tour guide has laminated slides for each stop, some historical (Victorian carriages in the park) and some personal (the West Highland terrier with an overbite belonging to one of his children). Anticipation slowly builds. "Only three more stops until we're at the cemetery . . . almost at the cemetery . . . this is the last pause before the cemetery!" Jonathan points to Richard, a Royal Parks employee who has the cemetery key in his fleece pocket. Richard is also responsible for other walks in Hyde Park, including "Mythical Trees" and "Bat Walk." "Don't lose that key, Richard!" Jonathan says. Thirty minutes later, with the cemetery but moments away, he says, "Richard, check you've still got that key!" Richard holds up an old silver key large enough for a door in a Christmas pantomime. "Good man!"

The key unlocks the melancholy; the cemetery is not a disappointment. It is compact and packed, the size of a tennis court. It began when the grief-stricken owners of a Maltese terrier named Cherry mentioned their loss to a friend named Mr. Winbridge, who also happened to be the keeper of Victoria Lodge at the edge of the park. He offered to bury the dog in his garden, and the first memorial arose: "Poor Cherry. Died April 28 1881." Word got around. Soon a Yorkshire terrier named Prince took up permanent residence close to Cherry, having been run over by a carriage nearby. The owners of Tar, Tubby and Jack the Dandy joined the gang, as did Joker, Boss and Curly. Then some cats died and they got in too. Mr. Winbridge did well out of this, charging £5 a pet, including a brief memorial service. The gravestone maker also dined well in those years.

All the dogs here are good dogs. There are no ne'er-do-wells, no biters. And no paupers either. These headstones are from the top of the Victorian pyramid, and most dogs did not enjoy these privileges either in life or death. Here, resting among the bluebells, are not necessarily the most loved, just the most permanently remembered, and their memory continues to affect us down the years. The memorial stones are the size allotted in human cemeteries for children, and a child may come here to find the gravestones to be the size of playhouses.

Many of the names on the headstones in Hyde Park are still popular today: Monty, Carlo, Jack. Others—Fido, Ruff, Rex—have achieved a status both mythical and ironic.

Others one hardly hears from these days, such as faithful old Scum. There's a Fritz too, and Fritz's son Balu, on the same gravestone, mysteriously immortalized as "poisoned by a cruel Swiss." By the start of the First World War the garden was full, although a few dogs have sneaked in since by special dispensation, the last of them Prince, a regimental mascot who died at age eleven in 1967. How was Prince remembered by his survivors? "He asked for so little," says his headstone, "and gave so much."*

Their ghosts live on in other cemeteries in other eras. At the Cimetière des Chiens in Paris lie more than forty thousand pets, including an early version (there were several) of the movie dog Rin Tin Tin. At Aspen Hill Memorial Park in Maryland one finds the wonderfully inscribed gravestone of a dog called Major: "Born a Dog, Died a Gentleman." Aspen Hill is the final home of more than fifty thousand pets, including hounds of stage and screen and several dogs owned by J. Edgar Hoover. The place began in the 1920s as a kennel and breeding ground, but now thousands of dogs and cats have memorials there, and quite a few goats, turtles, frogs and hamsters. It's the second-largest pet cemetery in the United States, after

---

* Early photographs show some gravestones that are no longer visible, and far more space allotted to each grave than is the case now. It's likely that the ground has seen more than three hundred burials, with some dogs and their headstones transported to the dog cemetery in Molesworth, Huntington. The Hyde Park cemetery was certainly not the first dog cemetery in England. The Duchess of York buried more than sixty dogs in the grounds of Oatlands House in Surrey around the turn of the nineteenth century (it's now a grand hotel, but the tiny gravestones remain), and Queen Victoria had her own smaller pets' burial ground at Windsor.

Hartsdale Canine Cemetery in Westchester County, New York, which, since its foundation in 1896, has buried more than seventy thousand pets, including those of many famous owners, such as Diana Ross and Mariah Carey. Your dog can join them, for the cemetery is still accepting new arrivals from the wealthy. At Hartsdale we find a familiar epitaph, this time for a collie:

OUR SYDNEY

DIED SEPT. 4, 1902

AGED 16 YRS.

BORN A DOG

LIVED LIKE A GENTLEMAN

DIED BELOVED.

In 2013, the anthropologist Stanley Brandes noted that the dogs buried at Hartsdale in the last two or three decades carried increasingly human names compared to those buried a half century before. Writing in *Names: A Journal of Onomastics*, he notes that the majority of dogs buried there were once called Laddie, Rex, Rags, Boogles, Trixie, Snap, Jaba "and similar names that are entirely uncharacteristic of human beings." Even in the 1980s, Champ, Happy and Spaghetti predominated, all names that no parent would give a child. But by 2013 one could already see a radical shift. "It is now very common to encounter inscriptions to dogs named Ronnie, Rebecca, Jasper, Marcello, Oliver, Fred, and Timothy."

The pet gravestone is a salve to us, a marker for our children, an emotional jolt to strangers. Others may prefer a less public or ostentatious tribute, or a less pressing sign of our own mortality. As John Galsworthy noted, "Not the least hard thing to bear when they go from us, these quiet friends, is that they carry away with them so many years of our lives . . ." Recalling his own dog, the novelist is flooded only with pleasant memories. "No stone stands over where he lies. It is on our hearts that his life is engraved."

HOW ELSE may a dog be immortalized? In the books we've already encountered, and in the paintings and cartoons we've already examined. And then there are public statues, cold and immovable with mythical legend attached, the myth increasing with each passing gaze.

To fully understand what may be achieved in this field we would do well to visit Tokyo, and two sculptures made seventy years apart. The sculptures are both of the same dog, and they tell the same story. Perhaps it's a story you know, so often has it been told, so highly do we value it. It shows dogs at their very best, and humans sometimes not at their very best.

The fable begins in 1923 with the birth of a cream-colored dog named Hachi. Hachi was a purebred Akita, and his owner lived in the Akita prefecture of Honshu. When Hachi was two months old, his owner heard of a professor at Tokyo Imperial University named Hidesaburo Ueno who was keen to have a friendly face to come home to every day after work. And so, because there were so many Akitas in Akita, and it was thought that one little one wouldn't be missed, Hachi journeyed for almost a day by train to meet his new friend. They were happy; it was an elegant exchange of love.

Dr. Ueno was a professor of agricultural engineering, and he traveled to work each day from his local railway station in the Tokyo district of Shibuya. Hachi would accompany him every morning, and be there at the station when the professor returned home. But less than two years into their friendship, in May 1925, tragedy struck both owner and dog. The professor had a fatal stroke at work, but no one told Hachi. No one! And so he stayed at Shibuya Station until nightfall, and returned every day in the hope of seeing the professor. Initially he survived on scraps and the comfort of strangers. Hachi must have wondered whether the professor had forgotten about him. Legend has it—and inevitably the legend is huge—that Hachi kept his faith for years: "Maybe today," he must have thought, "maybe today . . ."

But after a few months some of the locals and railway staff found Hachi a nuisance, and tried to discourage him from returning. He was getting in people's way at a very busy station; he was an unnecessary cause for concern. *Didn't he understand that his owner was dead?* In an effort to discourage him, there were tales of poor Hachi being beaten. Others took pity on him and continued to supply him with food, but the food was no salve for a broken heart.

Professor Ueno also had a human partner for the last years of his life, Yaeko Sakano, but they never married; their relationship was frowned upon by her family, and she did not adopt Hachi. After a while, to save him from abuse and confusion, Kikuzaburo Kobayashi, the professor's former gardener, made the dog his companion. They lived together in Tomigaya, and while their bond was beneficent, it wasn't love. Fortunately, the gardener lived close enough to Shibuya Station for Hachi to walk there every day, his hope undiminished.

In 1932, after Hachi had been waiting daily for seven years, a reporter from a national newspaper turned up and realized that something special was going on. Radio stations and other newspapers picked up the story—this dog was lovesick, this dog was sensational—and for weeks it was all anyone could talk about. This was not just a stray station dog, but an animal with unbound optimism! That we should all have such faith! Hachi became a symbol of innocence and loyalty, contributing something important to a nation still struggling to find its identity between the wars. And his fans added something in return: the suffix "kō," to denote affection and cuteness.

Hachikō didn't actually look very cute. Photos showed him stocky and cock-eared, and possessed of a certain screw-you attitude; the dog appeared to know his own mind. After his fame spread, people sent care packages (mostly meat) from every part of the Japanese archipelago, and so, at the same time that he was becoming the most revered dog in the world, he faced a struggle not also to become the fattest. There was so much food arriving at Shibuya Station that no stray dog within an area of five miles would ever go hungry.

How should one commemorate such a dog? The same way one commemorates humans: a statue in a public place.

Less common, perhaps, is the actual attendance at the unveiling ceremony of the figure being celebrated, but Hachikō was indeed on the red carpet at Shibuya Station in 1934, a year before he died. The bronze was by Teru Ando, who looks very sad in the one photo that exists of him. He made a figure so stout and immovable—Hachikō high on a plinth, his stance reminiscent of a sumo at a basho—that on its first appearance one could imagine it staying there high on its plinth well after the trains at Shibuya had been replaced by rocket ships.

Hachikō died from cancer a year after the unveiling at the age of eleven, and when they opened him up they found four metal chicken skewers in his stomach.

Hundreds of thousands of people pass his bronze statue each day at the northwest exit of Shibuya Station (the "Hachikō Exit"), and they touch his feet in the hope that his soul may live in theirs. The bronze by his feet has become smooth and lighter in shade from all the touching. However it is not the original they caress, but a replica made fourteen years later by Ando's son Takeshi, for the first Hachikō was melted down for ammunition during the Second World War.

No single dog has had more impact on a culture. Hachikō's fur has been preserved and stuffed and placed inside glass at the National Museum of Nature and Science in Tokyo, the place where you may also find exhibits on the early Japanese car industry and a prototype of the Sony Walkman. In Japan, to commemorate the dog, there are stamps and school musicals, and national days of celebration (in 1994, a restored recording of Hachikō barking was played to millions on the radio, and the country was reported to have stood still during the broadcast). There have been several films, most notably, in 2009, *Hachi: A Dog's Tale*, in which Richard Gere plays the professor. The professor's name has been changed from Hidesaburo Ueno to Parker Wilson, providing the first hint that the movie may not be as faithful as the dog. On the plus side, the film is a surefire tearjerker, and was directed Swedishly by Lasse Hallström (he's a dog man, all right: he also made *My Life as a Dog* and *A Dog's Purpose*).*

There are other important bronzes of the hound. The most joyous is Tsutomo Ueda's double statue of Professor Ueno greeting an enthusiastic Hachikō on the Hongo campus of his Tokyo university, unveiled on the eightieth

---

* Alas, Hachikō has endured other, less noble uses down the years. Like all dogs who transcend their mortality, Hachikō has become representational, a symbol of his nation's ambitions and fortitude. As an agent of imperialism, Hachikō and the Akita breed came to symbolize a proud cultural and military future; astonishingly, given the dog's rather hapless appearance and tender soul, the selfless commitment required of Japan's soldiers in a period of military aggression also took some of its lead from him. When the deaf-blind writer and activist Helen Keller received a gift of an Akita after visiting Japan in 1937, and it became the first Akita to enter the United States, the dog already had a name: Kamikaze-Go.

Hachikō also changed the landscape for all Japanese breeds. Through most of the nineteenth century, when pet ownership of dogs in Japan was limited to royal families or the extremely privileged, Western breeds were largely regarded as superior; indigenous breeds were either shunned or destroyed. But Hachikō irreversibly made a case for the Akita as a national treasure, and the breed remains the most popular native of Japan (followed by the Shiba Inu, the Shikoku, the Kai Ken and the Japanese spitz).

anniversary of the dog's death in 2015. This is a slightly idealized representation, and certainly a slimmer one. But it's public art to lift the public spirit, and is every person who passes it not obligated (and delighted!) to reimagine the tale that inspired it? Some stories do not diminish or waver over time: Why should the next generation be any less enthralled?

The story of canine loyalty is one we like to tell ourselves repeatedly. We have already seen Fido in Italy. Smoky the parachuting therapy dog also has a bronze to herself, in Cleveland, Ohio (she's sitting in a GI's helmet, and the inscription reads, "Smoky: Yorkie Doodle Dandy"). Then there is Balto in Central Park, the Siberian husky sled dog who in 1925 ensured the anti-diphtheria vaccine made it through the ice and snow to Nome, Alaska. And Jirō, another symbol of icy endurance, the shaggy black Sakhalin husky who survived for a year at the Shōwa research station in Antarctica in the 1950s after Japanese scientists abandoned him. And who among us can fail to shed a tear when chancing upon the memorial statue at a bleak roundabout in Kraków, a tribute to the hapless Polish mutt Dżok (pronounced "Jock"), who declined to leave the fatal spot where his master had a heart attack in his car in 1990?

Most infamously of all, we have Edinburgh's memorial fountain for Greyfriars Bobby, the Skye terrier who sat by his master's grave for fourteen years in the mid-nineteenth century. Experts have wondered whether there weren't actually several Bobbys, or Bobbies, and several owners, and whether those concerned with the well-being of the Greyfriars tourist trade weren't responsible for a bit of hokum over the years.*

But do not all dogs deserve a statue of their own? Or if not one each, then perhaps a universal shrine?

There is indeed such a shrine, and it stands outside the Johnson County Courthouse in Warrensburg, Missouri. Here a statue commemorates a foxhound named Old Drum, and it commemorates an oration delivered in his defense in 1869 by the lawyer George Graham Vest. Old Drum was shot by his owner's neighbor for straying onto his property, and the dog's owner sued for wrongful killing and compensation. Lawyer Vest appeared to ignore all the arguments in the case, and instead spoke to a higher nobility. Not all of his speech was transcribed, but an extract appears beneath Old Drum's memorial:

> Gentlemen of the jury, the best friend a man has in this world may turn against him and become his

---

* The news of the terrier was first reported in the *Ayrshire Express* in 1865, and the newspaper conceded that its story was already almost six years old. Its language was monumental and alliterative. "A terrier dog was found lying under a horizontal grave-stone in Old Greyfriars' grave-yard . . . The poor brute had evidently been there some days, and, although exhausted with hunger and thirst, viciously refused to be removed." Only occasionally could Bob, as he was originally called, be enticed away from the kirkyard. "During the inclement weather the year before last," the *Ayrshire Express* reported, "Sergeant Scott, of the Royal Engineers, one of Bob's best friends, got him coaxed into his house for a night or two." Though he soon returned to the gravestone, the dog swiftly learned of the sergeant's dinner hours and visited him when the hour was right.

> enemy. His son or daughter that he has reared with loving care may prove ungrateful. Those who are nearest and dearest to us, those whom we trust with our happiness and our good name, may become traitors to their faith. . . . The people who are prone to fall on their knees to do us honor when success is with us may be the first to throw the stone of malice when failure settles its cloud upon our heads. The one absolutely unselfish friend a man can have in this selfish world, the one that never deserts him and the one that never proves ungrateful or treacherous is his dog.

The speech swayed the jury. The owner was awarded the maximum reparation of $150, not that the money brought back his dog.* The statue went up outside the courthouse in 1958 after donations came in from forty states and six countries overseas. The memorial, like a dog, may not be wholly trusted: as the writer and sculptor Stephen Huneck once observed, you can trust a dog with your life, but maybe not your lunch; and anyone who has ever failed to give their dog an entire pack of treats at once will attest to what ingratitude means. But the statue is a reassuring thing to see—such a solid assertion of justice, such a victory for the dog. To me it

* Vest went on to become a U.S. senator. His speech further popularized the notion (and phraseology) of the dog being man's best friend, a concept permanently enshrined by Voltaire in his *Dictionnaire Philosophique* of 1764. It surely won't need saying that we should also always consider the dog to be woman's best friend too.

also says: our relationship with the dog is one of the finest things we've ever created, and we should be daily proud of this feat, based as it is on the principle of mutual trust and permanent allegiance.

A DOG represents grief built-in: we acquire our friends with the knowledge that we will one day come to mourn them.

Once, only poetry offered consolation, soft anthems for the untethered. Thomas Hardy, for example, adopts his own dog's melancholy air:

Do you look for me at times,
When the hour for walking chimes,
On that grassy path that climbs
Up the hill?

(From "Dead 'Wessex' the Dog to the Household")

Wordsworth writes of the tears he shed as he buried his dog Little Music beneath an oak tree, but also gratitude for a final release:

For thou hadst lived till everything that cheers
In thee had yielded to the weight of years;
Extreme old age had wasted thee away,
And left thee but a glimmering of the day.

(From "Tribute," 1805)

Given the subject matter, and given the medium, one would expect a certain amount of doggerel; Sydney Smith doesn't disappoint:

Here lies poor Nick, an honest creature,
Of faithful, gentle, courteous nature;
A parlour pet unspoil'd by favour,
A pattern of good dog behaviour.
Without a wish, without a dream
Beyond his home and friends at Cheam . . .

(From "Here Lies Poor Nick," date uncertain)

The most personal may also be the most universal:

The rugs lie smooth; the curtains are not torn,
I haven't missed a shoe or rag today.
The house is dreadfully still, until I wish
I heard four feet come pitpat down the hall.

The soft moist nose that pushed against my hand
The paw that touched me to demand its wish,
The pleading lively eye, the plaintive bark–
What sweet annoyances they now would seem!

(From "To a Dog," writer unknown)

But today we have a new path to consolation, a shared sadness online. Where go humans, so go dogs: a beautiful memorial for a dog may help others similarly stricken, or may help us prepare for a bereavement of our own. The website dogquotations.com carries many such tributes, and I was surprised how easily my eyes welled up when reading about dogs I never knew. But as we mourn our dogs, we mourn our own impermanence alongside them.

Our Beautiful Girl Maggie

By Roger and Deb

We miss you with all our heart.

The snoring and whimpering when you were dreaming. We miss the little twitch of your crooked ear, and your lip catching on your tooth. The clacking of your nails on the wood floor. Your kisses were so special, even though they were stinky kisses.

You were a very precious and special dog. Except that you were our sweetheart, and we hate that you are not home with us. But we eventually will be with you again.

You did not deserve to be uncomfortable, so that is why we had to let you go. We both held you, until you went into a peaceful sleep. We had 13 beautiful years with you, and treasured every moment we shared with you.

We hope and pray that you knew how much we loved you. All we can hope is that you felt that love.

There will NEVER be another dog like you, Maggs, and I hope you enjoyed your life with us.

Rest In Peace,

our Dolly

At a Loss for Words

By Neil

Okie, you were an awesome dog! Always loyal and loving to your family. Strangers didn't get you, but that's ok. We loved and embraced your quirks!

I still feel so much grief over your loss . . . I know we had 13 great years but I just didn't think we were close to the end. That still breaks my heart.

I really wish that you had a chance to say goodbye to the girls. I felt so bad that they were several states away when you were taken from us!

I miss you so much. The house is so quiet without you! I have cried so many times, trying to process all of this, but I'm glad that you are not in pain anymore! #foreverinmyheart

My Jeep

by Jeep's Momma (Indiana)

When I heard you needed a home, I thought, "I'm so busy all the time. I don't need a dog again right now."

Then the day I came to meet you (because something just kept tugging at me, telling me I should at least meet you) you came bouncing up with the big oversized ball you had popped in your mouth . . .

The first day I had to leave you at home alone I thought, "He is so big, he is probably going to eat my whole house while I'm out. What am I doing? I don't need a dog right now."

You didn't eat the whole house that day or any day after. You never had any bad behavior (ever). It's almost like you knew, "She doesn't think she needs a dog."

As the years passed we grew so close (sometimes I think people in our lives were jealous of the devotion we shared), but you just got me. At times I thought you understood me better than most of those humans.

You have been my comedian with your little silly antics anytime I needed to laugh, my furry tissue when I needed to cry, my walking buddy, my car ride buddy, and, at times, the reason I didn't give up.

You, my friend, were never a dog in my eyes. It was clear looking in your eyes that you were an old soul more than likely sent here to point mine on the right path. You did that job well.

There are no words strong enough to thank you for allowing me to be your person, a job I hope you feel I did well. For 11 years, you were my best friend.

Over the last 4 days, as hard as I have tried, I can't stop the tears. I can't remember what life without you in it was like, but I guess I have to somehow figure that out now.

Until we meet again. You will be missed more than words can express.

AND WHAT would the dogs say, given half a chance?

In December 1940, near the end of his life, the Dalmatian Silverdene Emblem O'Neill, known to his friends as Blemie, wrote his last will and testament. Compositional help came from the playwright Eugene O'Neill, who saw nothing exceptional in his dog taking the lead toward the end of his life. The will was intended to ease the pain of O'Neill's wife, Carlotta, once Blemie passed on.

"I have little in the way of material things to leave," Blemie began. "Dogs are wiser than men. They do not set great store upon things. They do not waste their days hoarding property. They do not ruin their sleep worrying about how to keep the objects they have, and to obtain the objects they have not. There is nothing of value I have to bequeath except my love and faith."

Blemie found it painful to think that his death should cause his owners any misery. But the time had surely come to ease his own: he had cancer and his faculties were all now failing. "Even my sense of smell fails me so that a rabbit could be right under my nose and I might not know." Dogs do not fear death as humans do, their dog wrote, but rather viewed the future with optimism. He envisioned a Paradise "where each blissful hour is mealtime; where in long evenings there are a million fireplaces with logs forever burning, and one curls oneself up and blinks into the flames and nods and dreams, remembering the old brave days on earth, and the love of one's Master and Mistress."

Blemie had one final request. "I have heard my Mistress say, 'When Blemie dies we must never have another dog. I love him so much I could never love another one.' Now I would ask her, for love of me, to have another. It would be a poor tribute to my memory never to have a dog again." Predictably perhaps, Blemie recommended another

Dalmatian, to whom he bequeathed his collar, leash, overcoat and raincoat.*

And one final word of farewell. "Whenever you visit my grave, say to yourselves with regret but also with happiness in your hearts at the remembrance of my long happy life with you: 'Here lies one who loved us and whom we loved.' No matter how deep my sleep I shall hear you, and not all the power of death can keep my spirit from wagging a grateful tail."

Inspired by these words, my Labrador was keen to add a few valedictory comments of his own. Ludo is now an aging gentleman; he is well into his thirteenth year, and although he's still in good health, I'm aware that a gradual decline may not be so distant. I've known a Labrador retriever that lived to seventeen, but twelve or thirteen is generally considered a good life span; my last dog, Chewy, died of cancer when he was ten. We'll all miss Ludo terribly when he dies—the soft padding around the house, the rejuvenating daily walks, the warm snoring at night. What would his own farewell contain? Optimism, I'm sure, and a similar desire to Blemie's that we'll not mourn him excessively. I think he'd leave us a set of rules to light the way for dogs yet to come, and if a couple of them sound like bumper stickers, remember that as far as dogs are concerned, bumper stickers are as good as it gets, height-wise.

---

* These were indeed all his possessions, made to order at Hermès when the O'Neills lived in Paris in the late 1920s. The fancy outfits we afford our dogs today are not so novel.

THE BEST THINGS IN LIFE AREN'T THINGS.

IF YOU SEE A BALL, THROW IT.

IF YOU GO TO THE CINEMA IN THE EVENING, WE WILL OFTEN GO ON YOUR SOFA IN THE EVENING.

PET INSURANCE IS ALWAYS WORTH IT IN THE LONG RUN.

JOIN:
DARWIN'S ARK
THE CAMPAIGN FOR LOWER KITCHEN COUNTERS

ASPIRE TO INTELLIGENCE. STAND UP FOR WHAT'S RIGHT. DON'T BELITTLE ANYONE, DOG OR HUMAN, PEDIGREE OR MUTT. TREAT US AS YOUR EQUAL AND WE SHALL NEVER LOOK DOWN ON YOU.

WHEN YOU'RE ALONE AND LIFE IS MAKING YOU LONELY, YOU CAN ALWAYS GO DOWNTOWN.

But what if you really can't bear to say goodbye to your dog?

If this is the case, you now have another option in the form of canine genetic engineering—cloning—the replacement of natural selection with intelligent design. This is the ultimate symbol of human mastery over the dog. It's a sad victory, I think, an inevitable fault line in our relationship, a corruption of human decency.

Not so long ago, the Victorians surely believed that a pure, carefully bred pointer or setter was all a fine dog could be. And not long after that (relative to the history of all dogs), the labradoodle set the pace. But now the future is here, and what was once a prophetic horror story has become a standard technological procedure.

At the Sooam Biotech Research Foundation on the outskirts of Seoul, South Korea, Dr. Hwang Woo-suk will make you a fresh dog for about $100,000. The Sooam Biotech website makes it all sound straightforward. To paraphrase:

> IMMEDIATELY AFTER DEATH, WRAP THE ENTIRE BODY IN WET BATHING TOWELS.
>
> PLACE IT IN THE FRIDGE (NOT THE FREEZER) TO KEEP IT COOL. YOU THEN HAVE APPROXIMATELY FIVE DAYS TO SUCCESSFULLY EXTRACT AND SECURE LIVE CELLS.
>
> THE GENETIC SAMPLES SHOULD BE EXTRACTED AND PRESERVED BY A VET, AND THESE THEN NEED TO BE FLOWN IN PERSONAL HAND LUGGAGE TO THE LABORATORY IN SEOUL, WHERE DR. HWANG, A MAN WHO WAS PREVIOUSLY DISMISSED FROM HIS POST AT SEOUL NATIONAL UNIVERSITY FOR FALSIFYING STEM-CELL RESEARCH, REPORTS A 40 PERCENT SUCCESS RATE.

This procedure, and its results, have drawn disapproval from all but the wealthiest and most fanatical. But Barbra Streisand is both, and in March 2018 she wrote in the *New York Times* that she felt only pride in her devotion to her dogs. "It was easier to let Sammie go if I knew I could keep some part of her alive, something that came from her DNA," she explained. In fact, Sammie, who was a straight-haired Coton de Tuléar, lives on—desperately and vaguely—in three dogs (another died not long after birth), all cloned in a laboratory run by ViaGen Pets in Cedar Park, Texas. ViaGen Pets describes itself as "America's Pet Cloning Experts," producing its first successfully cloned dog, a Jack Russell terrier, in 2016. Subsequent business was only ticking along slowly until Streisand sang her eulogy, after which interest peaked. In 2019, *Bark* magazine quoted president Blake Russell as saying, "We're weekly producing cloned puppies at rates we're very happy with."

Should you wish to visit Cedar Park in person, ViaGen has a useful map on its website showing that it is right next to Costco Wholesale. Those who are tempted, and have dogs who are still alive, may sign up for Genetic Preservation, a $1,600 option that ensures the freezing of genetic material to be used as soon as occasion demands. This, of course, raises the possibility of not waiting for your dog to die at all, but producing a companion litter to fuss over while the original can still enjoy it. The FAQ section contains an exceptionally unnerving inquiry: "How do I know that my genetically preserved or cloned pet is authentic?"

But the work in Texas and Seoul does not "bring back your dog," as Streisand and many others hope, but a dog made from its cells. And because of what we know from the hugely complex makeup of a dog's genetic structure from the Broad Institute and elsewhere, the cloned dog may have a completely different character and susceptibility to new disease. And as if one needed further evidence of the fragile unpredictability of life, a cloned dog's appearance and health may be additionally influenced by the health and hormones not only of the original cells from which it's cloned, but also those of its surrogate mother.*

---

* The procedure is also clearly the cause of cruelty and distress to the surrogate dogs that carry the clone. When the journalist Richard Lloyd Parry visited Dr. Hwang's laboratory for *The Times* toward the end of 2018, he witnessed the delivery of Puppy 1,192, a cloned English Bulldog. This dog would have to wait several months in laboratory kennels, alongside some other 150 clones, undergoing many developmental tests, before its new owner could meet quarantine requirements and take it home.

This is a wholly unnatural disruption of the way we have lived with dogs for thousands of years, and raises a host of ethical questions. Two of the most prominent are: How to legislate the progress of technology and science for a species unable to legislate such progress itself? And, after all this time, do we not have a greater responsibility and loyalty to dogs than this? Where is the good news about our relationship? Where does our best future lie?

14.

Discover Dogs!

In some ways, the best future for our dogs lies online. Here we may access all the understanding quite unavailable to a previous generation. We may chat with multiple breeders, obtain best practice information from the Kennel Club, join general dog forums and specific breed forums, and have almost any question answered by an expert. We may learn about the health issues of our dogs, and how to improve them. And we participate in the largest citizen science project ever conducted with dogs, a vast genetic study you will find at darwinsark.org.

I signed up, and received the following enthusiastic email:

> Welcome! Thank you for joining Darwin's Ark. We're excited to begin this journey of scientific discovery with you.
>
> At its heart, the Darwin's Ark project is a study of how genetics shape the heath and behavior of animals and humans. In addition

> to the survey questions, we're sequencing thousands of samples of canine DNA to help us discover the roadmap to shared health. Your dog's information will be a part of this incredible body of knowledge . . .
>
> We are super excited to have 22,230 dogs on the Ark as of today! Together, we can collect the critical data needed to secure a better future for both our animals and ourselves.

Darwin's Ark is not part of a futuristic film or novel, and it is not a parody. It is a genuine and valuable thing, and any dog may join, pedigree or mutt, good boy or not—in fact, the more dogs the better. The surveys in this particular type of science are diverse, covering many elements of a dog's behavior and attitude to life in general. The themes range from "Play," "Communication" and "Socialization with Humans" to "Coping with the Unfamiliar," "Canine Eccentricities" and "Socialization with Humans II." The site automatically personalizes its questions. Assuming you register your dog as Ronald, you will soon receive such inquiries as, "How much time in a normal day does Ronald spend false-digging (scratching at floor/carpet/etc.)?" And "How much time in a normal day does Ronald spend showing trance-like behavior?"

The site also offers something else, the ability to sequence your dog's DNA in one of the most advanced research labs in the world. This is now a standard procedure, as common for dogs as it is for humans, with a similar variability in accuracy. There are several testing options, your choice determined by your impatience and curiosity. Option number one is the basic Free Kit Level. After completing at least ten short surveys you'll be sent a free kit to swab inside a dog's cheek, you send it away, and after a while you'll get some results about your dog's breed and ancestry. But "a while" will certainly be several months, and possibly a couple of years.

Then there is the Explorer Level at $149. Your dog's DNA swab will be sent to a lab for small-batch testing, which means that you should get your results in between 90 and 120 days.

Can't wait that long and feeling flush? Welcome to the Trailblazer Level. This costs $2,199, or $1,999 after ten surveys, and not only is it fast, it is also thorough and useful. "Be the foundation upon which knowledge is built," the Darwin's Ark researchers say. "With your donation, your dog will join an elite group of fewer than 1,000 dogs worldwide with a complete DNA sequence, informing scientists globally in the advancement and study of canine and human health." Trailblazer Level samples will receive immediate sequencing at "30 × coverage," which means each of the 2.4 billion bases (A, C, G, T) is sequenced thirty times to ensure exceptionally high accuracy of data. Commercial dog DNA tests offer between 1,800 to 200,000 generic markers, but Darwin's Ark's technology offers over 4 million genetic markers, offering a far deeper delve into a dog's ancestry. You'll also receive "a personalized certificate of scientific contribution" signed by real-life scientists.

Let's take the example of a peculiar little creature called Gunther. Gunther CentralPerk is a combination dog. He was born in February 2018, and it is safe to assume that his owner once enjoyed watching *Friends.* The people he meets say he looks like a fox, because of his big ears, or like a deer, due to his long legs. His owner has another description for him, employing that peculiar language that perhaps only dog owners know how to appreciate: "This little goober is a love bug." What makes him special? For one thing, he's cunning. "If he wants a toy that his sis Remi has," his owner reports, "he will ask to go outside, wait for me to get up and Remi to follow me, then race back to where she left the toy and steal it."

The other thing that makes him special is his ancestry. Before signing up at Darwin's Ark, Gunther's owner sent a swab from the dog's mouth for a DNA sampling at another, less exacting company. The results suggested he was 25 percent miniature pinscher, 12.5 percent Chihuahua, 12.5 percent Yorkshire terrier and 12.5 percent boxer. The rest of him, some 37.5 percent, was classified as the breed "unknown."

The more exacting analysis at Darwin's Ark judged Gunther 30.3 percent miniature pinscher, 6.4 percent Chihuahua, 4.7 percent Pomeranian, 3.1 percent toy poodle, 3 percent American cocker spaniel, 2.7 percent pug and 49.8 percent "unknown." If you ever needed proof that dogs can be complicated things, here it is, in an algorithmic maelstrom of genetic coding. Where did his trace of pug come from? Where did his eighth of boxer and Yorkie go? They came and went in a demonstration of scientific fallibility. What we don't know about dogs—such as the date they began to emerge from the DNA of wolves—is still inestimable, even if progress is continuous.

Whatever the hell Gunther is—and he is nothing if not lovely, with an ability to jump six feet in the air when he's happy—he isn't passing it on: he's been neutered. But he has already made his modest contribution to the baffling world of dogs.

In terms of "breed" we can most accurately call him a village dog, which, given that about 80 percent of all dogs are believed to be village dogs, makes Gunther part of the most comprehensive and complicated breeding program on earth. The fact that Gunther's owner reports that Gunther is also "a natural stand-up paddle boarder" is unlikely to help in our meaningful quest to understand where dogs came from, or why they behave like they do, but it does suggest yet another milestone in our quest to make dogs behave just like us.

DARWIN'S ARK is an experiment, and its aim is monumental. It represents true citizen science, where all the citizens are dogs. The premise—and it's nothing if not brilliantly clever—is to link DNA analysis with a dog's behavior. The behavioral surveys provide one set of answers (subjective, quirky), and the DNA another (comprehensive, precise), and at some point it may be possible to show how changes in DNA lead to changes in behavior.

But it is early days. Even with more than 4 million genetic markers, the Trailblazer Level is not yet able to supply the one thing you might really want—an indication of your dog's predictive health. Gunther will not learn, for example, about the presence or possibility of breed-specific disease, or how likely it is that a mutt made up of miniature pinscher and Pomeranian may contract one of the ailments common to those breeds. "We feel strongly that only well-validated science be included in our reports to dog owners," the Darwin's Ark researchers explain. "We also feel strongly that health tests that are offered for dogs should meet the same quality standards as those required in human medicine. To date, they do not meet those standards." One aim of Darwin's Ark is to bring forward the date when they will.

The woman who runs Darwin's Ark (or actually, for the time being, its first and sole offshoot, Darwin's Dogs) is Dr. Elinor Karlsson, the director of the Vertebrate Genomics Group at the Broad Institute in Cambridge, part of MIT and Harvard. She describes herself on her Twitter page as "Immigrant. Scientist and artist," and she described herself to me as "really a cat person" (her cats are called Darwin and Lacey, the latter short for Ada Lovelace, the mathematician and computer pioneer). But dogs are where the action is.

"Dogs are an amazing genetic model," she told me when we met in December 2018. "And the science is moving so fast. Everything that we thought we knew we're now thinking about differently. Everything we were doing two years ago, we're not doing it that way anymore." She is interested in the genetic changes that enabled the transformation from wolf to dog, not least the changes that enabled mutually beneficial human interaction.

"The example I use is retrieving a ball," Karlsson says. "Retrieving is actually a variation on the wolf hunting sequence. The pattern for wolves taking down large prey has a few steps. They see it; they lower their head and stalk it;

they chase it; they do a grab-bite nip at the heels to take it down; and then they deliver a bleeding bite to the jugular." Karlsson says that our domestication of dogs ensured that, over many centuries, humans removed the "bite to the jugular" element out of dogs pretty early on. "It's not that it's not there, but it's pretty hard for dogs to enter into it. So dogs get really, really into the chase part and the grab, but they don't go any further. It's still a complex behavior. There are plenty of purebred retrievers that don't retrieve. It's not like just turning on one gene."

Darwin's Dogs has already gathered genetic reference points for more than one hundred different breeds, a list that includes such rarities as the Norwegian elkhound and the Belgian Tervuren. But they love mutts too, Karlsson tells me as we drive to Boston South Station at the end of her day. (Part of our interview was conducted in her office, part on a tour of the Broad sequencing laboratories and part in her new Tesla Model 3, the Trailblazer Level of electric cars that monitors what it calls "your driving DNA," before sending it back to the manufacturer.) "We're currently examining to what extent mixed-breed dogs who have more ancestry from a pure-breed with a particular behavior trait have also inherited more signs of that behavior. It works beautifully for retrieving, tying in very well as to how much Labrador retriever ancestry is in the mutt." She now hopes to apply this to more complicated personality traits. "And we mustn't forget that 80 percent of dogs in the world are village dogs. 'Domestication' does not mean a cute purebred puppy on a couch . . ."

Darwin's Dogs takes up only one part of Elinor

Karlsson's professional life. She is also energized by something she had begun working on as a junior researcher in 2003 and which reached completion two years later: the first complete sequencing of a dog's DNA. In the first week of December 2005, when *Nature* appeared with seven dogs on its cover, the news made headlines worldwide.

Why seven dogs? Six (among them a collie, Dalmatian, Labrador and spaniel) were seen from behind, looking up at a framed black-and-white photograph of James Watson and Francis Crick demonstrating the structure of the double helix in 1953. The seventh dog, a female boxer named Tasha, had been Photoshopped into the Watson/Crick picture, as if she'd been there all along. Tasha was now the star—the dog whose blood had been extracted to achieve the sequence.*

"There were all these cameras and reporters there, and then they brought a dog in," Dr. Karlsson remembers of the press conference to announce the initial findings. "And as soon as you get a dog in the room, no one takes much notice of the humans anymore." The principal researchers, Kerstin Lindblad-Toh, Elaine Ostrander and Eric Lander, explained the methods and significance of their work. Using Tasha's genome as a compass, they had then sampled the genomes of ten other breeds as well as a gray wolf and a coyote. They found 2.5 million individual genetic differences among breeds (called single nucleotide

---

* Tasha was six when her blood was first taken. She belonged to somebody who was in the collaborative research network, so it was easy to obtain the owner's consent. After the announcement, the owner was keen to shield themselves and Tasha from further publicity.

polymorphisms), which could be used as signposts to locate the genetic contributions to physical and behavioral traits, and help understand the huge number of diseases afflicting purebred dogs.

But there was also another motivation for their work. "Dogs and humans share many diseases, including cancer, epilepsy and diabetes," Dr. Karlsson explained at the time. "By directly comparing the disease genes found in dogs to genes in humans, discoveries made in dogs can benefit human medicine." Much of her subsequent work would then focus on locating that very small part of the genome that has some influence on dog behavior or disease (because most of the genome doesn't). By the time of the announcement of the full dog sequence, several locations for a specific canine disease had already been located on their chromosomes, including the preponderance of narcolepsy in Dobermans, but there were many more ongoing searches—for hip dysplasia, deafness, motor neuron disease, cataracts, heart defects and OCD among them. In a dog, OCD manifests itself as excessive licking or repeatedly chasing a tail.

The press conference then moved on to questions and answers. The work had taken almost three years and cost many millions of dollars. The work was far more thorough than Craig Venter's attempt two years before to extract a DNA sequence from his poodle Shadow, an enterprise that covered only 80 percent of the dog's genome and left many important gaps. There had been some prior debate as to whether a chimpanzee should be the next mammal to be sequenced after the mouse (completed in 2002) and the human (completed in 2003). But the dog won out, not only because of the diseases shared with humans (chimps have that attribute too), but also because of the large amount of knowledge that could be gained from examining the differences in so many breeds (despite the huge differences in their physical appearance, a dog is a dog is a dog: every dog has a genome that is 99.85 percent similar to every other dog's). Tasha the boxer was chosen for two reasons: tests suggested that the boxer was more inbred than the other dogs they were looking at (which might make her susceptibility to specific disease and other traits more easily identifiable), and her white front and solid brown back indicated she was heterozygous, possessing two different coat-color gene pairs on a specific chromosome; researchers could then try to isolate this mutation.*

But this was the relatively simple stuff. "I fear that people's expectations got raised a little high," Dr. Karlsson told me. "There was a feeling among some people that cancer in dogs was going to behave like coat color, and that we

---

* As opposed to being homozygous, where the pairs (or alleles) on each gene are identical. The change in a genome when comparing one dog with another is extremely similar to the change when comparing one human with another. They are 99.9 percent the same, with about one change in every one thousand genomic bases. But although humans and dogs share a similar biology, canine inbreeding means that the number of dogs one needs to conduct experiments is only a fraction of the number one needs of humans. So where around twenty thousand humans would be required to map breast cancer, in dogs perhaps one hundred English springer spaniels with breast cancer can be compared with one hundred control dogs for a similar sort of analysis.

were going to cure cancer in dogs."* There is unlikely to be a single gene that causes a bone cancer that, for example, greyhounds are particularly susceptible to, and finding a gene that correlates to something is a long way from understanding how it works or how to counteract it; the key is not just how one gene reacts on its own, but how they interact with each other.

But several significant scientific papers have already emerged from the work begun with Tasha and her ten comparative breeds (Tasha herself died of lymphoma at the age of twelve). The amount of genetic information now available, and the increased speed and reduced cost of processing it, has advanced our understanding of early domestication, not least one project suggesting that dogs in the Americas were domesticated twice—once thousands of years ago from Arctic dogs that accompanied human migrations from Asia, and once again a little more than five hundred years ago after the initial contact with Europeans. In terms of disease, there have been important developments in the understanding of the pathways of oncological and autoimmune mutations common in Dobermans, Shar Peis, German shepherds and Siberian huskies.

This is a tantalizingly new world for humans and their

---

* Coat color in a dog is determined by a combination of eight specific genes controlling variations in two basic pigments. One pigment (eumelanin) is a default black, which may be genetically diluted into a gray or brown, while the other pigment (phaeomelanin) is an Irish Setter red that may dilute into a cream, yellow or golden shade.

dogs, united by forces quite unimaginable when our relationship with dogs began.

IN OCTOBER 2019, I went to Discover Dogs at ExCeL in London's Docklands, an experience also quite unimaginable when our relationship with dogs began. This is London's biggest dog event, and consists of 35,000 people speed-dating for their perfect match. There are almost two hundred different breeds on show, each with their owners in their own little booth, and both dogs and owners will be pleased to answer any questions and exchange phone numbers.

As at Crufts, the booths are arranged alphabetically, so popular breeds such as the King Charles Spaniel compete for your attention with the Kooikerhondje (a small Dutch spaniel, sociable to the point of frisky, a great retriever, slightly feathery coat, known to Henry VIII), the Korean Jindo (highly intelligent if a bit of a trickster, bolts of energy, loyal like glue) and the Lagotto Romagnolo (small Italian water-loving gundog, popular in the Renaissance, matted hair coat, superb truffle hunter). The owners of these dogs are proselytizers for their breeds, and they have traveled from all over the U.K. and Europe to take part at their own expense. I was pleased to see the Canaan dog still thriving after those appearances on rocks from 500 B.C. in the Shuwaymis and Jubbah regions of Saudi Arabia. I was delighted that the giant Hungarian Kuvasz and the small Hungarian Pumi were proving so popular after so many years in the rare-breed wilderness, and that the Entlebucher mountain dog, the Estrela mountain dog, the Bavarian mountain dog and the Pyrenean mountain dog had not only come down from their respective mountains and seemed to be perfectly healthy amid the denser vapors of London's Docklands, but also seemed to be the best of friends when they met.

There was a lovely atmosphere in this vast warehouse, a shameless lovefest that manifested itself in an endless round of sighs from adults and children. This culminated in Cuddle Corner, a candy-colored area where visitors reclined on beanbags and were licked and pawed by puppies as excited as they were. Had the event not been indoors the sky surely would have been all rainbows.

A few days before I went to Discover Dogs I talked to Caz Brixton, a successful breeder of French bulldogs. She had a warning for me, and it was the same advice I'd heard from several other breeders I'd talked to that year: be very wary when buying a new dog, because "it's a minefield out there"

Brixton lives in a vicarage in Newlyn, near Penzance in Cornwall, with her partner and five French bulldogs (or "Frenchies," as the dog fancy like to call them). "Beau's on the left with her gray muzzle," she explained, as I examined a panting tableau on the striped rug on the floor of her home office. "She's the mother of Jasmine, the one with her tongue out, the puppy from her second litter, and then at the back is Diesel, her stud dog, who I'm still waiting to use." I observed that Diesel looked ready for action. "Yes," she said. There's also Bella, who is out of the picture somewhere, and Hugo, who was the one who started it all about five years ago.

"I was not in a very good place after my dad died," Brixton told me. "My partner is a trainer, and one of the people he trains had a French bulldog, and one Saturday morning I met the dog and he filled all of my emotional gaps." Brixton was then in her late forties and hadn't lived with a dog since she was a child, but she resolved to find a Frenchie from a local breeder. She came up short, reading online about a lot of health issues associated with the breed, and waited almost a year before she found a dog she liked—Hugo—who lived in Canterbury. When Hugo displayed signs of loneliness and became excessively clingy, Brixton decided to find him a mate, but again had trouble locating one she liked the look of. This time she had a ten-hour drive to the Midlands, and was shocked by what she found there.

"When I arrived you could tell it was a house that didn't have any dogs in it, and the mum and dad weren't there as the breeder had promised. I could see they'd had her ears taped up to make them stronger. It became a rescue case—I couldn't leave her there, so I paid for her and left." Beau cost her £4,000.

French bulldogs have their origins in the ancient Phoenician world, but the modern breed, popularized in the nineteenth century, is a cross between a toy English bulldog and a French street dog with a skill for ratting. The dogs were commonly employed in bear baiting, but swiftly became acceptable as edgy lapdogs, a position they retain. The breed is currently in huge demand, and a new owner can expect to pay upward of $2,500.* Caz Brixton described them as "quirky, a bit like little kids, and you get loads back." A standard Kennel Club description will note their muscular, heavy build and smooth coat, and their outsize bat ears; they often look as if they may be picking up signals from another universe.

In Newlyn, Caz Brixton runs her own consultancy business, enabling schools and other institutions to meet their legal child-protection requirements. Because she is self-employed, she finds it relatively easy to scale down her work when she needs to nurture pups for their new homes. She is meticulous with her screening for potential new owners, sending each a detailed letter outlining what each party should expect from the other. "The puppies are crate trained (just in case owners wish to use them)," part of the letter explains. "They are aware and respond to key words, they know and come to their name, are socialised and feeding independently and I take pride in the substantial puppy packs I provide for them." The current packs include detailed breed information, a certificate of microchipping, first vaccination and Kennel Club Registration, a squeaky toy, antiseptic ear wipes for the dog and antiseptic handwash for the owner, poo bags and dispenser, a scented

* On a recent visit to New York I saw a sign in the window of the West Village pet store Citipups announcing "We have Frenchies!" as if they were the latest Prada bag. In 2018 they were the most registered breed in New York, and the fourth most popular in the U.S. In the U.K., the Kennel Club reported that in the first quarter of 2018, there were 8,403 registered French bulldogs versus 7,409 Labradors, the first time in twenty-eight years that the Labrador wasn't in the top spot. The ones on display in Manhattan were a red fawn merle color and had been reduced from $4,295 to $2,995; "30% Off," the sign said, "Take Me Home Christmas Specials."

puppy fleece from her pen to remind the pup of her mother and ease the transition, among many other things.

When we spoke, Brixton was preparing for her dogs' third litter, with nine pups already assigned on a waiting list. But not all French bulldogs are the same, or elicit the same enthusiasm from potential owners. "People prefer blue-tone puppies," Brixton explained, as opposed to ones with a brindle or fawn coat. This has created a problem for her, as her eager stud Diesel doesn't carry the blue gene, and she has been obliged to hire another stud to mate with Jasmine in the hope of either 'blue, blue fawn, blue tan or blue sable with fawn points." Her letter to owners assures them "Jasmine carries amazing DNA . . . she has also been health screened and her breathing is amazing!!"

So much of a dog's future prospects—its health, its sociability with other dogs, its success as a family pet—stems from the first ten weeks of its life, and the cost for this caring start is considerable, ranging from £2,200 to £2,600 per pup; Brixton says she will let a dog go at the lower end if she feels it's a perfect fit for a family, and if a new owner has no intention to breed from the dog. Alas, she has seen the dog too often become a commodity. "I'm still on a very steep learning curve with people around French bulldogs and now I don't trust any of them," she told me. "I sold one recently to a person who said they were the love of their life and all that, and next thing I know they've sold him to America."

YOU CANNOT buy a dog at Discover Dogs, but you can buy almost anything else. The commercial area contains many products just for humans—dog-walking jackets and hats, a huge confectionary stall—and many products just for dogs: toys, many types of food, quick-dry wraparound furry coats. But then, as at Crufts, there are stalls where it is increasingly difficult to determine if the items are for humans or dogs, such as the Dog Bowl Deli selling food disguised as pies and tiered birthday cakes, and Different Dog recipes, where samples of turkey fricassee and lamb hot pot (the latter with chia seeds, kelp, turmeric, banana, salmon oil, and added vitamins A and D, zinc, iron, manganese and copper) are freshly cooked in a wok by an enthusiastic woman in an apron. And on the way out a human could sign up for dog yoga, also known as doga, in which a regular class is enhanced by a beautiful if lawless array of hounds of all ages and creeds walking around your mat as you stretch downward and enlighten.

When I got home I wondered again about a dog's purpose. At Discover Dogs I had strolled among the Dogues de Bordeaux, the Canadian Eskimo dogs and the Norwegian elkhounds, and it seemed no longer necessary to determine whether humans existed for the dogs' happiness or they existed for ours. We just coexisted, and we were muddling along just fine, unaffected (at least for now) by the specter of cloning and unsavory breeding practices, impressed but unbothered by the clothes that would keep them dry and the diets that would keep them trim. For here was everything a dog required: a vast amount of people giving them their love.

But after the show I wanted to return to a place where dogs roamed free and seldom encountered chia seeds, and where the only fur coats were their own. I had always liked a much-quoted observation from the novelist Milan Kundera: "To sit with a dog on a hillside on a glorious afternoon is to be back in Eden, where doing nothing was not boring—it was peace." So I went for a walk on Hampstead Heath with Ludo, and I savored how much simple fun he was having among the scents of all the other dogs who had walked these fields before. I said hello to other dog owners, and we exchanged information about breeds, allergies and the weather. With our dogs, we were all walking around with books of great stories. People who would rarely say hello if I was walking on my own immediately relaxed into conversation; I am seen as more reliable, less threatening, softer, with a dog. Dog ownership, I realized again, binds us not just to one being but to a wider world, a world of responsibility and sociability, a purposeful community.

A little farther on my walk with Ludo we passed the spot where we had said a tearful goodbye to our previous family Labrador thirteen years before. We scattered Chewy's ashes in a puddle; he adored puddles so much, the muddier the better. He hated being hosed down afterward, but he never equated getting dirty with the water jet that inevitably followed: he lived for the moment and for pleasure. And now Ludo was also having a wonderful adventure, sniffing at every tree and raising his leg at most of them. A dog slows the passage of time. I remembered a poem Siegfried Sassoon wrote during the war: not a war poem, but a few lines published in *Country Life* in 1941 about walking with his thick-coated Dandie Dinmont terrier, a gift from his friend Rosamond Lehman, a tranquil respite from conflict.

Who's this—alone with stone and sky?
It's only my old dog and I—
It's only him; it's only me;
Alone with stone and grass and tree.

What share we most—we two together?
Smells, and awareness of the weather.
What is it makes us more than dust?
My trust in him; in me his trust.

Here's anyhow one decent thing
That life to man and dog can bring;
One decent thing, remultiplied
Till earth's last dog and man have died.

What was true in 1941 remains so in our own turbulent times: the bond between us and our dogs may be the one of the few decent things we bring to the world. Our relationship is based on trust, and there is no falsity in our love. That is why we are here for them, and they for us.

As we walked on, I realized that Ludo doesn't care much for his origin story or his DNA analysis, and he spends very little time contemplating which of his far-flung relatives might one day win Best in Show. He is still best at being the thing he was ten thousand years ago, despite all that has befallen his species, despite everything we have done to make him more like us. He is best at being a dog. He gets very excited about the prospect of lunch, or any food, really, and he usually comes running when I call him, and we're always impossibly happy when we're together.

ACKNOWLEDGMENTS

One of the most rewarding elements of writing this book has been the huge outpouring of enthusiasm from dog owners who wanted to share their stories. I am grateful to them all, but a few deserve special mention.

The idea for this book began in an email exchange with my editor, Jenny Lord, who knew of my lifelong love of dogs and thought it was about time I wrote about them. Her editorial advice and judgment have been erudite and immensely beneficial to the entire project, but above all she ensured this book was fun to make. Her colleagues at Weidenfeld have been a pleasure to work with, and I would like to thank everyone who has contributed to the book, in particular Ellie Freedman, Sarah Fortune, Cathy Dunn, Steve Marking, Hannah Cox and Susan Howe.

Seán Costello copyedited with his customary precision and finesse, and it's no accident I return to him with my manuscript of each new book.

My agent, Rosemary Scoular at United Agents, and her assistant, Natalia Lucas, have been as supportive and ingenious as ever.

The staff at the London Library possess no limits to their truffling talents.

I am delighted to work with my new American editor, Nick Amphlett, and all at William Morrow, and my old Italian friends at Ponte alla Grazie.

Andrew Bud read the manuscript with his usual enthusiastic sticklerism for grammar and wayward clauses.

The thriving presence of Daunt Books has provided its familiar solace, not least my local branch managed by Mary and her team in South End Green. If I worked anywhere but my desk I'd like to work there.

Thanks to Ben and Jake Garfield, Charlie and Jack Drew, Inês Afonso, Izzy O'Bryen, Bobbye Fermie and Morgan Smith for keeping me supplied with enough dog stories and stupid pet jokes to fill the Golders Green Hippodrome. More, please.

Lisa, Steve, Louis and Noah Gershon have always been great supporters of my work, and have provided the best second home Ludo could have wished for.

And not forgetting all those who shared their thoughts about this book in particular and dogs in general: Daniel Pick, Mark Ellingham, Don Guttenplan, Ralph and Patricia Kanter, Diane Samuels, Tessa Shaw and Ink@84, Stephen Grosz, Plum Fraiser, Georgie Ferraro, Suzanne Hodgart, Catherine Kanter, Hal and Georgia Kanter Condou, Abby Hollick, Dan, Oscar, Lenny and Joseph Benoliel.

And finally to my wife, Justine Kanter, who truly understands the value of dogs and love.

FURTHER READING

We adore dogs, so we write about them. We value their role in our lives, and the part they have played in human history, so we write about them. We should only be grateful that they don't write about us nearly so often.

The "further reading" selection that follows is the most pertinent to my subject and I hope the most rewarding. You will find all manner of canine diversions here; you could do worse than regard these books as great walks with dogs of all breeds. Because I too have a dog to walk I have not included all the academic journals consulted; the majority can be accessed via jstor.org, where, if you enter the simplest search for "Dogs," you will be rewarded with 247,817 articles and book excerpts (as of the beginning of April 2020).

Ackerley, J. R., *My Dog Tulip* (London: Secker & Warburg, 1956)

Ash, Edward C., *Dogs: Their History and Development* (London: E. Benn, 1927)

Bailey, Paul, *A Dog's Life* (London: Hamish Hamilton, 2003)

Baker, Steve, *Picturing the Beast: Animals, Identity and Representation* (Manchester: Manchester University Press, 1993)

Berns, Gregory, *How Dogs Love Us* (Boston & New York: Houghton Mifflin Harcourt, 2013)

Bicknell, Ethel E., ed, *Praise of the Dog: An Anthology* (London: Grant Richards, 1902)

Big New Yorker Book of Dogs (New York: Cornerstone, 2012)

Bondeson, Jan, *Amazing Dogs: A Cabinet of Canine Curiosities* (Gloucestershire: Amberley, 2011)

Borjesson, Gary, *Willing Dogs & Reluctant Masters* (Philadelphia: Paul Dry Books, 2012)

Bradford, Arthur, *Dogwalker* (New York: Knopf, 2001)

Budiansky, Stephen, *The Truth About Dogs* (London: Weidenfeld and Nicolson, 2001)

Caius, John, *Of Englishe Dogges, the Diversities, the Names, the Natures and the Properties: a Short Treatise Written in Latine* (1576; repr. London: A. Bradley, 1880)

Campbell, Clare, *Bonzo's War* (London: Constable & Robinson, 2014)

Caras, Roger, *A Celebration of Dogs* (New York: Time Books, 1992)

Carr, Neil, *Domestic Animals and Leisure* (Houndmills, Hampshire: Palgrave Macmillan, 2015)

Chance, Michael, *Our Princesses and Their Dogs* (London: John Murray, 1936)

Coppinger, Raymond and Feinstein, Mark, *How Dogs Work* (Chicago: University of Chicago Press, 2015)

Coren, Stanley, *How to Speak Dog: Mastering the Art of Dog-Human Communication* (New York: Simon & Schuster, 2001)

Coren, Stanley, *The Pawprints of History: Dogs and the Course of Human Events* (New York: Free Press, 2002)

Dalziel, Hugh, *British Dogs* (London: Gill, 1888)

Dean, Emily, *Everybody Died, So I Got a Dog* (London: Hodder & Stoughton, 2019)

Derr, Mark, *Dog's Best Friend: Annals of the Dog-Human Relationship* (New York: Henry Holt, 1997)

Dodd, Lynley, *Hairy Maclary from Donaldson's Dairy* (London: Puffin, 2002)

Don, Monty, *Nigel: My Family and Other Dogs* (London: Two Roads, 2016)

Dubbs, Chris, *Space Dogs: Pioneers of Space Travel* (New York: Writer's Showcase, 2003)

Eastman, P. D., *Go, Dog. Go!* (London: Random House, 1961)

Fogle, Ben, *Labrador: The Story of the World's Favourite Dog* (London: William Collins, 2015)

Fogle, Bruce, *The Dog's Mind* (London: Pelham Books, 1990)

Garber, Marjorie, *Dog Love* (London: Hamish Hamilton, 1997)

Grandin, Temple and Johnson, Catherine, *Animals in Translation* (London: Bloomsbury, 2005)

Grandin, Temple and Johnson, Catherine, *Animals Make Us Human* (Boston and New York: Houghton Mifflin Harcourt, 2009)

Gray, Beryl, *The Dog in the Dickensian Imagination* (Farnham: Ashgate, 2014)

Green, Susie, *Dogs in Art* (London: Reaktion Books, 2019)

Grenier, Roger, *The Difficulty of Being a Dog* (Chicago, University of Chicago Press, 2000)

Grossman, Loyd, *The Dog's Tale: A History of Man's Best Friend* (London: BBC Books, 1993)

Haddon, Celia, *Faithful to the End* (New York: St. Martin's Press, 1991)

Hall, Bernard J. and Foss, Valerie, *Treasures of the Kennel Club* (London: Kennel Club, 2000)

Hausman, Gerald and Loretta, *The Mythology of Dogs* (New York: St. Martin's Press, 1997)

Hawtree, Christopher, *The Literary Companion to Dogs* (London: Sinclair-Stevenson, 1993)

Homans, John, *What's a Dog For?* (London: Penguin, 2012)

Horowitz, Alexandra, *Inside of a Dog: What Dogs See, Smell and Know* (New York: Scribner, 2012)

Horowitz, Alexandra, *Our Dogs, Ourselves* (London: Simon & Schuster UK, 2019)

Hughes, Jimmy Quentin, *Who Cares Who Wins* (Liverpool: Charico Press, 1998)

Jackson, Frank, *Faithful Friends: Dogs in Life and Literature* (London: Robinson, 1997)

Jenkins, Garry, *A Home of their Own: The Heartwarming 150-year History of Battersea Dogs & Cats Home* (London: Bantam Press, 2010)

Jesse, Edward, *Anecdotes of Dogs* (London: R. Bentley, 1846)

Junor, Penny, *All the Queen's Corgis* (London: Hodder & Stoughton, 2018)

Kean, Hilda, *The Great Cat and Dog Massacre* (Chicago: University of Chicago Press, 2018)

Kennel Club's Illustrated Breed Standards, 4th ed. (London: Ebury Press, 2011)

Lane, Charles Henry, *All About Dogs: A Book for Doggy People* (London and New York: J. Lane, 1900)

Laybourn, Keith, *Going to the Dogs: A History of Greyhound Racing in Britain, 1926–2017* (Manchester: Manchester University Press, 2019)

Lemish, Michael G., *War Dogs: Canines in Combat* (Washington, D.C.: Brassey's, 1996)

London, Jack, *The Call of the Wild* (New York: Bantam, 1963)

Lorenz, Konrad, *Man Meets Dog* (London: Methuen, 1954)

Lucas, E. V., *If Dogs Could Write* (London: Methuen, 1929)

Masson, Jeffrey M., *Dogs Never Lie About Love* (London: Cape, 1997)

McConnell, Patricia, B., *For the Love of a Dog* (New York: Ballantine, 2005)

Menzies, Lucy, *The First Friend: An Anthology of the Friendship of Man and Dog Compiled from the Literature of All Ages 1400 B.C.–1921 A.D.* (London: Allen & Unwin, 1922)

Merwin, Henry Childs, *Dogs and Men* (Boston and New York: Houghton Mifflin, 1910)

Miklósi, Ádám, *The Dog: A Natural History* (Brighton: Ivy Press, 2018)

Morey, Darcy, *Dogs: Domestication and Development of a Social Bond* (Cambridge: Cambridge University Press, 2010)

Pemberton, Neil and Worboys, Michael, *Mad Dogs and Englishmen: Rabies in Britain, 1830–2000* (Houndmills, Hampshire: Palgrave Macmillan, 2007)

Pierce, Jessica, *The Last Walk: Reflections on Our Pets at the End of Their Lives* (Chicago: University of Chicago Press, 2012)

Ritvo, Harriet, *The Animal Estate: The English and Other Creatures in the Victorian Age* (Cambridge: Harvard University Press, 1987)

Rogers, Katharine M., *First Friend: A History of Dogs and Humans* (New York: St. Martin's Press, 2005)

Rosenblum, Robert, *The Dog in Art from Rococo to Postmodernism* (New York: Abrams, 1988)

Sackville-West, Vita, *Faces: Profiles of Dogs* (1961; repr. London: Daunt Books, 2019)

Schaffer, Michael, *One Nation Under Dog* (New York: Henry Holt, 2009)

Sheldrake, Rupert, *Dogs That Know When Their Owners Are Coming Home* (London: Hutchinson, 1999)

Simon, Joan, *William Wegman: Funney/Strange* (New Haven: Yale University Press, 2006)

Skabelund, Aaron Herald, *Empire of Dogs: Canines, Japan, and the Making of the Modern Imperial World* (Ithaca, NY: London: Cornell University Press, 2011)

Smith, Arthur Croxton, *Dogs since 1900* (London: A. Dakers, 1950)

Sorenson, John, and Matsuoka, Atsuko, eds., *Dog's Best Friend? Rethinking Canid-Human Relations* (Montreal & Kingston: McGill-Queen's University Press, 2019)

Spicer, Kate, *Lost Dog* (London: Ebury Press, 2019)

Thomas, Elizabeth Marshall, *The Hidden Life of Dogs* (Boston: Houghton Mifflin, 1993)

Thompson, Laura, *The Dogs: A Personal History of Greyhound Racing* (London: High Stakes, 2003)

Thurston, Mary Elizabeth, *The Lost History of the Canine Race: Our 15,000-year Love Affair with Dogs* (Kansas City: Andrews and McMeel, 1996)

Trew, Cecil G., *The Story of the Dog and his Uses to Mankind* (London: Methuen & Co. 1940)

Turkina, Olesya, *Soviet Space Dogs* (London: Murray and Sorrell Fuel, 2014)

Wang, Xiaoming and Tedford, Richard H., *Dogs: Their Fossil Relatives and Evolutionary History* (New York: Columbia University Press, 2008)

Watson, James, *The Dog Book* (London: W. Heinemann, 1906).

Woolf, Virginia, *Flush: A Biography* (London: Hogarth Press, 1933)

Worboys, Michael, Strange, Julie-Marie and Pemberton, Neil, *The Invention of the Modern Dog: Breed and Blood in Victorian Britain* (Baltimore: Johns Hopkins University Press, 2018)

Youatt, William, *The Dog* (London: Charles Knight and Co., 1845)

PHOTO CREDITS FOR INSERT

p. 1 Hans Christian Adam

p. 2 *top:* Getty/*The Washington Post*
bottom: Alamy/Melba Photo Agency

p. 3 *top:* Shutterstock/khd
bottom: Courtesy of the author

p. 4 *top:* Alamy/Keith Corrigan
bottom: Alamy/dpa Picture-Alliance

p. 5 © Elliott Erwitt/Magnum Photos

p. 6 *top:* Virginia Woolf
bottom: Alamy/Photo 12

p. 7 *top:* Beinecke Rare Book and Manuscript Library
bottom: Courtesy of the author

p. 8 *top:* Alamy/Interfoto
middle: Alamy/baypixx
bottom: © The Paul Kaye Collection / Mary Evans Picture Library

p. 9 *top:* Getty/Media Production
middle: Alamy/MBI
bottom: Alamy/Sputnik

p. 10 *top:* Alamy/Chronicle
bottom: Getty/Topical Press Agency

p. 11 Courtesy of the author

PHOTO CREDITS FOR INSERT

p. 12 *top:* Getty/Popperfoto
bottom: Getty/The Asahi Shimbun

p. 13 Getty/Lisa Sheridan

p. 14 *top:* Alamy/Tim Jones
bottom: Getty/Kazuhiro Nogi

p. 15 *top:* Courtesy of the author
middle: Alamy/CPA Media Pte Ltd
bottom: Getty/Manuel Velasco

p. 16 *top:* Courtesy of the author
bottom: Courtesy of the author

INDEX

INDEX

POSTSCRIPT

Ludo passed away just as this paperback was going to press in mid-July 2021, a month short of his fourteenth birthday. He was bravely enduring various ailments, but when his back legs finally gave way we decided to save him from further suffering. He was a magnificent dog, and he gave so much to everyone who met him. I shall miss him very much, and I hope this book sends him on his way with the love he deserves.